THE ORIGIN OF
DISCRETE PARTICLES

SERIES ON KNOTS AND EVERYTHING

Editor-in-charge: Louis H. Kauffman *(Univ. of Illinois, Chicago)*

The Series on Knots and Everything: is a book series polarized around the theory of knots. Volume 1 in the series is Louis H Kauffman's Knots and Physics.

One purpose of this series is to continue the exploration of many of the themes indicated in Volume 1. These themes reach out beyond knot theory into physics, mathematics, logic, linguistics, philosophy, biology and practical experience. All of these outreaches have relations with knot theory when knot theory is regarded as a pivot or meeting place for apparently separate ideas. Knots act as such a pivotal place. We do not fully understand why this is so. The series represents stages in the exploration of this nexus.

Details of the titles in this series to date give a picture of the enterprise.

*The complete list of the published volumes in the series, can also be found at
http://www.worldscibooks.com/series/skae_series

K&E Series on Knots and Everything — Vol. 42

THE ORIGIN OF DISCRETE PARTICLES

T. Bastin • C. W. Kilmister

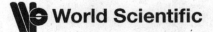
World Scientific

NEW JERSEY · LONDON · SINGAPORE · BEIJING · SHANGHAI · HONG KONG · TAIPEI · CHENNAI

Published by

World Scientific Publishing Co. Pte. Ltd.

5 Toh Tuck Link, Singapore 596224

USA office: 27 Warren Street, Suite 401-402, Hackensack, NJ 07601

UK office: 57 Shelton Street, Covent Garden, London WC2H 9HE

Library of Congress Cataloging-in-Publication Data
Bastin, Ted; Kilmister, Clive
   The Origin of Discrete Particles / by Ted Bastin & Clive Kilmister.
    p. cm. -- (Series on knots and everything ; v. 42)
   Includes bibliographical references and index.
   ISBN-13: 978-981-4261-67-8 (hardcover : alk. paper)
   ISBN-10: 981-4261-67-X (hardcover : alk. paper)
   1. Particles (Nuclear physics)--History. 2. Quantum physics--History. 3. Mathematical
physics--History. 4. Science--Philosophy--History. I. Title.
   QC793.16.B37 2009
   530.15--dc22

                                                                  2009013402

British Library Cataloguing-in-Publication Data
A catalogue record for this book is available from the British Library.

Printed in Singapore.

Foreword

During the twentieth century the concept of particle underwent a profound change. An electron differs from the particle of Newton in having discrete properties. All electrons have identical mass, charge and spin. The charge and spin of the electron and other elementary particles are logical properties which the particles either have or not. There are a number of other such logical properties. Mass is a more obscure matter but is logical none the less. This book attempts to show why such a change has come about. As a historical matter these investigations did not start from this problem. They began from trying to give physical interpretation to a curious algebraic construction of A.F.Parker-Rhodes. This construction gave rise to a graded algebra having exactly four levels with which were associated the numbers 3, 10, 137 and 10^{38}. The rough correspondence of the last two with the inverses of the electromagnetic and gravitational coupling constants and the restriction to four levels suggested a physical connection. It transpired that a very abstract account of the discrete process by which knowledge of the universe is gained provides a reason for an algebraic structure more general than Parker-Rhodes' but having his as a special case. The fact that the more general structure then modifies 137 to a value agreeing with the measured value of the inverse fine-structure constant to better than one part in ten million confirms its correctness. The discrete nature of elementary particles follows from the essentially discrete nature of the process. This explanation is logically prior to one invoking Planck's constant, since assuming the existence of Planck's constant is equivalent to assuming the discreteness. The fact that the algebraic structure contains a substructure agreeing with that of the quarks is a further confirmation of its correctness. Much more work is needed to evaluate all the implications for physics but it is hoped that this book will serve to inspire it.

Acknowledgments

The authors wish to thank many members of
the Alternative Natural Philosophy Association (ANPA)
for discussions of this work over the years, and in particular
Pierre Noyes and John Amson for unfailing helpfulness
even at times when their own views were different.

Contents

Chapter 1

Introduction

The method for physics put forward in this book is *constructive*. One may alternatively say that it is a *process* theory. In it one does not imagine the space and time continua as existing outside the theory and giving us a universe of objects already distributed in these continua to start from. All this has to be constructed.

Any construction of this kind must necessarily start from discrete quantities — numbers in fact. However remote that may seem from the classical basis of physics, we argue in this book that it is the place at which the physicists concerned with the high energy physics of the particles in fact have to start. The book constitutes a discussion from the process point of view of the evolution of the concept of *particle*. In high energy physics, new knowledge about particles has to be expressed in terms of properties like those familiar from classical physics even though they are not classical particles.

To make this clearer we begin by specifying what is meant by a classical particle. That concept has changed over time but by, say, the mid-eighteenth century a particle had one *primary* property, its mass, and two *secondary* properties, its position and its momentum. Change in position as time elapsed then produced a path. The other primary properties quoted by Newton (Opticks) of roundness and hardness were no longer necessary. As electromagnetism developed a second primary property, charge, was added. It would have been possible to introduce another secondary property, intrinsic spin, which could have been useful in the analysis of ball games, for example billiards, but as a matter of history, the notion of a rigid sphere was used instead to carry spin. The advent of relativity produced no great conceptual change by itself. It is true that relativity allowed a convenient description of the spinning particle but the stimulus for this came from quantum mechanics.

Quantum mechanics called for major changes in the concept of particle. The two secondary properties, position and momentum, were restricted by the uncertainty principle. In practice, the momentum became more important and the position increasingly disregarded so that the path became uncertain. The two primary properties, charge and mass, and the secondary one, spin, were taken over though only with important modification. The charge could have only three possible values including zero, a fundamental unit or its negative. A given particle has always the same charge. The mass could have a range of values, but again all particles of one kind had always the same mass. Finally, the spin became a third primary property whose value was some multiple of a fundamental unit.

The new knowledge that was coming in about particles in high energy physics has to be expressed by new primary properties. These are represented by descriptors which are called 'quantum numbers'. Descriptors are things that either exist or do not exist in any particular case. (For example some particles have charge though the neutron does not.) Conventionally they are given the values 0, 1 or digits reducible to 0, 1. The rules for assigning these quantum numbers capture their conservation laws in particle interactions. The first step from these descriptors toward the notion of a particle is the interaction strength. This is no longer an all-or-none property but has one definite value which characterizes a given particle process. An important investigation is always necessary to relate this value with the value of the mass, since mass will have a range of values even though it is not continuously variable.

The interaction strength is moreover a *number* that comes from counting of processes and only at a further remove should it be thought of as a field strength. The classical notion of a field is, in any case, inappropriate when the position of the particle is uncertain. Underlying the interaction strengths is a set of coupling constants that are dimensionless.

In our theory we find the origins of the coupling constants in a study of the building up of structures from interactions with an unknown background rather similar to the vacuum of high-energy particle physics. Construction does not go on into ever-increasing complexity but can be reversed or started all over again, so that there is a dynamic process at work.

This book is therefore an essay in the conceptual foundations of physics. Its purpose is to introduce what we shall call a *combinatorial* approach.

A process approach is necessarily combinatorial but the opposite is not the case. To clarify this consider a combinatorial approach in quite a different sphere, that of plane geometry. The fundamental entity there is the *point*. A pair of points is *defined* as a line. Two such pairs then define a fifth point. In normal parlance this would be the intersection of the two lines. Appropriate rules governing the derived entities then lead to the construction of projective geometry. (Euclidean geometry would be another matter.) In contrast the usual view would be of a continuum containing all the points of the plane and all their joins. These would all be on an equal footing so that it can only be an accident in what order they appear.

In physics our method will be to view physical theory under the aspect of a particular point of view which is combinatorial in character. In the course of the book questions will be asked and discussed which may have a long history, but which are not seen as live issues at the present because of the special philosophical stance of present-day physicists. However for combinatorial physics they are very much alive.

The idea which underlies combinatorial physics is that of *process*. The most fundamental knowledge that we can have is of step-by-step unfolding of things; so in a sequence. This is the kind of knowledge we have of quantum processes, and that fact becomes specially evident in the experimental techniques of high-energy physics. Such a process is necessarily combinatorial but not conversely. The example of plane geometry above exhibits a combinatorial theory which is not a process one.

The contrasting view, which has been the main guide of physics in the past, is of a background of physical things whose spatial relationships are on an equal footing in the sense that it can only be an accident in what order we happen across them. To put it another way, when the sequence or process is fundamental then we have to specify the steps by which we get to a particular point, whereas in the conventional picture we imagine we are free to move about in the space in which phenomena occur, without its being necessary to be explicit about it.

Suggestions for fundamental revision of the conceptual framework of physics are unlikely to engage the attention of physicists unless they bring about major improvements in the technical understanding of physics — particularly in new explanations and calculations of experimental results. The strong position of the combinatorial theory is that it has been used to deduce some experimental quantities that have not been deduced from more conventional theory. The closeness to the experimental values of these deductions makes it very unlikely that their success is fortuitous. Moreover

it would be agreed on all sides that the experimental quantities in question
play a sufficiently important part in physics to warrant attention being paid
to any theory that claims to calculate them.

Chapter 2

Combinatorial space

Contents

A hierachical theory presented in Chapters 6 and 7 is prefigured. In it, contact with the experimental world is through elementary interactions which appear as the strengths of elementary particles. Space is combinatorial and not absolute — or existing beforehand. Leibniz was the pioneer of this combinatorial approach. Computing models provide the connectivity of this relational space (this term for it used by Leibniz).

In this book we work with a space that is built up of elementary interactions. The first point of contact with the experimental world is the set of interaction strengths of fundamental particles and which give rise to their fields. We therefore postulate a combinatorial space; not a continuous or absolute space that exists independently. We have to be more explicit about what we mean by a combinatorial physical theory. Combinatorial physics is physics in which the mathematical relations are combinatorial, and combinatorial mathematics is mathematics in which we study the ways in which symbols are combined. The term 'combinatorial' is often defined ostensively by giving examples. Ryser[1] gives the following examples: interest in magic squares, study of permutations and (indeed) of combinations, finite projective geometry and problems connected with covering of spaces with arrangements of shapes in circumstances in which the space is divided into a finite number of sections and there is a convention which enables us to decide unambiguously whether or not a given section is covered. Such conventions exist, for example, for the use of chess boards and most other gaming boards, and the conventions enable us to decide all questions of the relation of combinatorial structures to physically defined spaces. Ryser[1] sums up his exemplification thus: 'Combinatorial mathematics cuts across the many subdivisions of mathematics, and this makes a formal definition difficult. But by and large it is concerned with the study of the arrangements of elements into sets.'

The first person to have seen a very profound difference between the combinatorial method and traditional thinking was Leibniz. The term 'combinatorial', used in this context, originates with Leibniz's "Dissertate de arte combinatoria", and according to Polya[2] the originator saw that difference in a very striking form. ... "Leibniz planned more and more applications of his combinatorial art or 'combinatorics': to coding and decoding, to games, to mortality tables, to the combination of observations. Also he widened more and more the scope of the subject. He sometimes regards combinatorics as one half of a general Art of Invention; this half should deal with synthesis, whereas the other half should deal with analysis. Combinatorics should deal, he says in another passage, with the same and the different, the similar and the dissimilar, the absolute and the relative, whereas ordinary mathematics deals with the one and the many, the great and the small, the whole and the part."

The general connexion between combinatorial mathematics and the impulse behind computing science has been widely remarked. Computer programmes are constructed in formal steps with information stored in (binary) patterns. It is the *connectivity* of the computer programme which is its vital characteristic, and we need to see this as an aspect of the 'ars combinatoria'. Weyl[3] says, "Modern computing machines translate our insight into the combinatorial structure of mathematics into practice by mechanical and electronic devices." When one is subjecting oneself to the discipline of the computer programming, which in its essentials is arranging the mathematical operations in sequential order so that each brings the next appropriate one in its train without intervention on the part of the mathematician, then (Weyl is saying) one has, through exercise of that discipline, displayed that insight.

Arguing this way, one comes to feel that in spite of all the modern sophistication about the impact of computing on the foundations of mathematics, not enough attention has been paid to the vast difference between the classical idea of the mathematician and that required by the computer revolution. This difference shows up clearly when it comes to making mathematical models. The mathematician makes up his mind about how to get from point to point in a mathematical structure on the basis of some purpose that he imagines to be directing his efforts, whereas, by contrast a computer person insists on having these instructions incorporated in the program.

What we abandon in making this change is the automatic freedom to consider a problem from several points of view and involving several mathematical techniques which have no coherence but which we take as simultaneously relevant. At this point the idea of space has changed in an important way. We expect a kind of coherence to these points of view and mathematical techniques because we associate them all to the same point of space. The combinatorial approach, by contrast, has to construct a *process* that plays the part of the composite picture of what is happening at a point of space. If we let a computing model stand for the combinatorial process, then we have to view every process from a preferred viewpoint which is conveyed by whatever is at the moment in the central processor. The possibility of parallel or concurrent computing does not change this argument and arises at a less fundamental stage: If there are several computers going, then insofar as we can speak of them as one model they are one computing system, even though the connexions between them may be randomized.

Can Leibniz' view of space fit his combinatorial method? In the first place Leibniz' space is essentially *relational*. So is time. There are *things*, and each pair of these things has a relationship together that we speak of as spatial and he does not suppose that there is a really existing scheme of all these relationships that exists independently of, and before the consideration of, the things. If it is possible to attach a number to characterize each relationship then we may call that the *distance* between those two things, though Leibniz does not tell us how that is to be done.

What are the things? The only answer given by Leibniz requires us to look into his all-embracing theory of the monads. The worldview of Leibniz — seen in relation to the Newtonian legacy of modern atomism — is well described by Whitehead[4]. "He" (Newton) "held the most simple-minded version of the Lucretian doctrine of the void, the most simple-minded version of the Lucretian doctrine of material atoms, and the most simple-minded version of Law imposed by Divine decree. His only approach to any metaphysical penetration arises from his way of connecting the Void with the sensorium of the Divine Nature. ... The monads of Leibniz constitute another version of the atomic doctrine of the Universe. Leibniz was acutely conscious of ... the problem of the criticism of knowledge. Thus he approached the problem of cosmology from the subjective side, whereas

Lucretius and Newton approach it from the objective point of view. They implicitly ask the question, What does the world of atoms look like to an intellect surveying it? ... But Leibniz answered another question. He explained what it must be like to be an atom. Lucretius tells us what an atom looks like to others, and Leibniz tells us how an atom is feeling about itself. Leibniz wrestles with a difficulty which infects modern cosmologies, a difficulty which Plato, Aristotle, Lucretius, and Newton entirely ignore. "The modern outlook arises from the slow influence of Aristotle's logic, during a period of two thousand years. Keeping to the point of view derived from Aristotle's logic, if we ask for a complete account of a real particular thing in the physical world, the adequate answer is expressed in terms of a set of abstract characteristics, which are united into an individual togetherness which is the real thing in question."

For Leibniz, this way of handling the particle concept is the only one possible. There never is a case where (as in the modern correspondence principle) it gradually becomes legitimate to go over into a classical mode of thought with its absolute space. Whitehead is drawing attention to Leibniz' doctrine on matter: an object is defined by the bundle of attributes that it has. Such a definition should be congenial to particle physics. But it does raise a problem — that of continuing identity — which is neatly dodged in the Newtonian synthesis by appealing to a continuity in temporal and spatial positions. No such appeal is open to Leibniz, (nor, incidentally to us) and this leads him on to the doctrine of monads.

Whitehead goes on to say of Leibniz's position that it "is beautifully simple. But it entirely leaves out of account the interconnexions between real things. Each substantial thing is thus conceived as complete in itself, without reference to any other substantial thing. Such an account of the ultimate atoms or of the ultimate monads, or of the ultimate objects enjoying experience, renders an interconnected world of real individuals unintelligible. The universe is shivered into a multitude of disconnected substantial things, each thing in its own way exemplifying its private bundle of abstract characters which have found a common home in its own substantial individuality. But substantial thing cannot call unto substantial thing."

This is the famous problem of the windowless monads. The system itself is necessarily incomplete, and Leibniz recognized this by breaking the system to the extent of allowing one supreme monad to have windows toward the rest. As far as his writing for the world of his contemporaries was concerned he could make a virtue of this necessity by identifying the supreme monad with the agglomeration of ideas that made up the then current perception of God. Whitehead comments "such has been the long slow influence of Aristotelian logic upon cosmological theory. Leibniz was the first, and by far the greatest philosopher who both accepted the modern" (outlook) "and frankly faced its difficulty".

Leibniz' writings on combinatorics only became available at the beginning of the nineteenth century. It appears that Leibniz found his Ars Combinatoria difficult to get across to his contemporaries and so did not publicise it. Russell argues that in doing this Leibniz was disingenuous because he liked the approbation that he got from the theism that his monadology permitted in spite of the fact that the combinatorics made the monadology unnecessary if not fallacious. Russell's view is wrong whatever he may have thought about Leibniz' morality. If we come only to the matters that are of direct concern to this book we find that the rather thin logical positivism embraced by Russell would never have presented the deep problems of modern physics satisfactorily, and Leibniz' combinatorics would not help anyway. That criticism would apply too to other writers of a positivistic orientation such as Popper (who concerned himself more directly with quantum theory than did Russell). Of course this is not to say that the monadology could solve these problems.

Russell[5] concludes: "What I, for my part, think best in his theory of monads is his two kinds of space, one subjective, in the perceptions of each monad, and one objective, consisting of the points of view of the various monads."

Though we reject Russell's sharp separation of combinatorics from the monadology, with the combinatorics as the way forward, that separation does seem to reappear in our approach. The process theory that we develop is thoroughly combinatorial, and our use of it as a basis for physical theory was not contemplated by Leibniz at all. By itself the Ars Combinatoria seems to be a bit thin. True, it prefigured much of modern computing science, but Leibniz would have demanded much more physical contact than that. Why could he not have that? The answer must be that Leibniz was born in the wrong century, whereas Newton was born in the right one. Celestial mechanics played into Newton's hands. If we may imagine Leibniz having been aware of the facts of quantum theory things might have been very different. We can play with the idea of Leibniz being offered the relational space of relativity and the combinatorial aspect of the world exhibited by the quantum theory, and wonder just how far he would not have gone with such a marvellous springboard.

Russell quotes Couturat[6] with great approval as saying "Leibniz's metaphysics rests solely upon the principles of his Logic, and proceeds entirely from them. Perhaps the most revolutionary conclusion in the whole book" (Couturat's) "is, that the principle of reason for all its trappings of teleology and Divine goodness, means no more than that, in every true proposition, the predicate is contained in the subject, i.e., that all truths are analytic." It is hard now, in the light of the criticism of Wittgenstein, to put oneself in Russell's position in which subjects and predicates and so on were completely commonsense and self-explanatory ideas, rather than having all the difficulties of metaphysical concepts. One cannot imagine Leibniz falling into this trap of logicist absolutism. He knew that time was not ripe for his ars combinatoria because such a metaphysics needs a subject matter which the physics of his day did not provide.

We turn from this account of Leibniz' thought to the way in which the authors began to put into effect the combinatorial ideas.

In current physics every question that may be raised about spatial relationships is referred to measurements of lengths expressed by numbers. Those numbers will provide the experimental meaning. If, now, we wish to replace absolute space by sequential construction, then we have to start afresh. To explain how we could still significantly speak about the physical world without the metrical provision of numerical relations we spoke of a *theory-language*. In the simplest theory-language dimension numbers were found in algebraic considerations separate from, and more basic than the metrical relationships.

Classical physics uses a continuum background of space within which things have positions, and the changes in those positions give rise to a second continuum of time. These continua are imagined as perfectly smooth, perfectly homogeneous, infinitely divisible. They are modelled mathematically by the continuum of all real numbers as that was formalized by Dedekind and Cantor between 1870 and 1880. The irreducibly simplest entities in these continua of space and time are particles — idealized as single points whose position in space changes continuously with time — and fields. Fields are spread through space and have at each spatial point a magnitude which again varies smoothly both with time and with changes of the point at which the field is specified. Upon this basis, a scheme of dynamical concepts was elaborated which seemed to provide a complete description of the motion of things in the space continuum. The work of Newton was the cornerstone of this edifice. With its subsequent extension to electromagnetism, it compelled a feeling of universality: one felt one was in possession of the means to describe reality direct. Indeed this scheme became for the physicist the mathematical elaboration of commonsense and the automatic vehicle for his thought. It was reality.

The set of concepts that makes up the scheme have an interlocking and closed character that makes it unique in the history of thought. Other disciplines have aspired to this cohesion, but have only very partially been able to achieve it. Through its long elaboration classical mechanics has become like a language that we learn to use rather than like an exploration that may turn out right or may turn out wrong, though of course it started off like that. When we learned classical mechanics we found to our surprise that the definitions of the concepts chase themselves round in circles. The mass of a body is given a numerical value through the behaviour of that body under a given applied force. The force is specified numerically by the acceleration of the body, but only if we already know the mass, and so on. At the time of Newton, there was no such unanimity on the use of the concepts, and if one were asked to give a definition of one of them, one would refer to some experimental situation which made the use one was proposing plausible: one would have to argue for it. Now things are interestingly different, for the set of interlocking concepts defines its own appropriate application, and cannot conflict with experiment without some major revolution having taken place. This set of interlocking concepts, with its dimensional structure that we find independently of metrical space, is our 'theory-language'.

For the great majority of physicists there is no problem in understanding the success of the theory-language of classical physics: they do not have to worry how it comes about that the theory-language describes the world so automatically because there is no distinction between it and reality. In a way we are suffering from the enormous success of classical physics. Over three centuries its concepts have penetrated our common language about the world almost completely, and the resulting amalgam of theory and experience seen under the aspect of that theory has become commonsense; it is identified with reality; and has the corresponding incorrigibility.

It is a strange fact that this immunity of the deductive language of classical physics to experimental check has received little or no attention in contemporary writing on the philosophy of physics. The term 'theory-language' refers to a theory which has reached this stage of development. The authors tried early on to use the 'logic of facts' of Wittgenstein's Tractatus Logico-philosophicus, to describe it : "... a physical theory consists of propositions which may be thoughts, sentences written or spoken, or manipulations with bits of the physical world ... the experimental thing has meaning only as part of a theory. The theory may have different degrees of complexity, and there will be experimental procedures corresponding to each degree. Thus the theory is a kind of language, but experiments in the theory are the same language. One cannot use experiments in a complex language to criticise a simpler theoretical language."

The simplest theory-languages, which play an important part in this book, are algebraic structures with a single binary operation. The first of these will be shown below to be the quadratic group S of 3 elements a, b, c together with an identity e. In this group $a^2 = b^2 = c^2 = e$ and $ab = ba = c$. The hierarchy of such structures consists of levels which are direct products of S with itself.

Let us suppose that the non-metrical basis of theory can be expressed as a theory-language and ask how metrical relations appear. The answer arises from the prevalence of the dynamics of the particle. The history of physics from Galileo onwards has been the progressive displacement by the dynamics of the particle of all other forms of explanation. Descartes' vortices failed because it proved not to be possible to calculate with them, and 'to calculate' increasingly came to mean 'calculate using the dynamics of the particle'. It is true that the great nineteenth century inventions of electromagnetism and fluid mechanics, which depend upon the theory of fluid rotation, could be seen as a vindication of Descartes' vortices, but really this would be to stretch a point, for the mode of action which Descartes imagined was not through particles. That came in the nineteenth century theory when all fields came to be specified in terms of the behaviour of ideal 'test-particles'. We don't understand the mechanics of rigid bodies and such things as gyroscopic motion until we have satisfied ourselves that the special concepts which they introduce can be made explicable in terms of particle dynamics. In fluid mechanics one must start with the 'experience' of the particle moving with the fluid.

Whatever the usefulness of the idea of the theory-language may be in general, it is an important part of our story because it was our attempt to formalize the idea that led to study of levels of interaction strengths and from that to the finding of the numerical values of them.

In an absolute space interpreted metrically, it is impossible to ask the question 'what happens when the metrical description fails to work?' The theory-language idea could provide independence of metrical application. It could, as it were, be carried around to see where it would fit; where it would not fit; and why. With no such recourse there is an obvious logical difficulty, if not actually an absurdity, in asking for a theory that accounts for, or even describes, the limits of applicability of itself. You must have something like a grid on which it is mapped to do that, and classical mechanics does not envisage anything of that sort. Both the great innovatory theories of this century — the quantum theory and relativity — found themselves up against this difficulty, and for that reason early forms of both theories took the form of arbitrarily imposed restrictions or alterations to the existing theory which came into effect at certain scales. This made the failure to fit of the theory look like an additional piece of mechanics of the type that the theory dealt in, when the true origins of the limits in each case were really quite different. The incorporation of the constant h into the mathematics of the quantum theory, and the Lorentz contraction, respectively, were the points of application of this method in the two theories. In these ways it was possible to put off the recognition that familiar metrically based dynamical concepts no longer applied universally.

Subsequently the theories became more sophisticated. In particular the inevitability of non-locality is recognized. However current ways of speaking about non-locality still show a backward-looking stance. One expects normal locality and takes to special devices when it fails, instead of seeing a more general form of connectivity as the norm. Very similar remarks apply to the quantum limits. Discussion of how one can have a dynamics without a classical background space and time when one adopts a combinatorial point of view will occupy a large part of this book, and we can stay at the level of general comment here. At that level it may be worth pointing out one principle that makes limits to the applicability of familiar dynamics easier to understand. One is accustomed to have a description of the world that is derived from our everyday laboratory-scale experience and to extrapolate it to different magnitudes. The technique and its very language suggest nothing that is not scale-invariant, and therefore we are puzzled to find it failing to work at particular scales, and have no way of describing that failure. How are we to understand limits to extrapolation? One piece of the puzzle arises from the way we take it for granted that there are an indefinite number of quite distinct ways of making the observations upon which our physical descriptions are based, and that these give consistent information. However, as we extrapolate further in both the small and the large scales, this presumption gets further and further from being valid. In the cosmological case, so far from having an indefinite number of cross-referencing methods of observing, we are reduced to very few, and ultimately to one. Notoriously, all our information about the universe on the largest scale depends on an interpretation of the red shift of spectra, while at the quantum end we are using the very particles we are investigating to do the investigating.

A quite insightful way of looking at what is happening is to say that theory and observation are losing their separability. By 'separability' we mean the situation, usually taken for granted, of a body of experimental results which is then rationalized or simplified to a greater or lesser extent by one of a number of possible theories. The choice of theory is influenced by, amongst other things, the extent of the rationalization produced. Here by contrast one gets towards a state where one would not know how to present the results without presupposing one particular theory. None of this discussion is in any way to impugn the validity of the observational work for what it is, nor to cast doubt on the care and sophistication with which deductions from it are made. It is only that current thinking allows for only one category of experimental knowledge, whereas we are confronted with something like a sliding scale from a situation where consistency is assured to one where it has no meaning. The very notion of observation has undergone a change in logical character. If we recognize that the experimental evidence and the theory can no longer be separated it is easier to understand the limit to extrapolation. It is usually true that adequate critical standards are applied to interpretations of experimental evidence near the extrapolation limits, yet the absence of a way of speaking that gives the experimental evidence its proper status cannot but cause confusion and possibly misdirection of effort. Thus in speaking of the big bang, talk of "the first three minutes" will certainly mislead people into taking the time language literally, and it is by no means certain that none of them will be professionals. There is no language that contains within itself a 'health warning' that there is trouble if you move back and forth over time while assuming the usual meaning for all the physical concepts.

The small pieces of matter that we usually associate with the mechanics of Galileo and Newton are always presupposed. However they have never been made a true part of that picture in the sense that their stability and impenetrability could be explained within the picture. The distribution of matter in the particles is never explained in a way consistent with the representation of the spatial distributions of those particles. Perhaps that is not surprising for it would seem to involve a circular argument. The clash of continuum and discrete which will occupy us centrally, becomes evident.

With the coming of field theories, and particularly of classical electromagnetism, this explanatory limitation became harder to overlook. The profound perplexities that arose in the continuum theories with the

Michelson-Morley experiment, and then with the ultra-violet catastrophe were given answers that had the effect of overlooking the problem about explanatory limitation for the time being. Discreteness was given an exact mathematical form by Planck, which seemed to be forced by experiment even though no one could justify it: it looked as though things were on the move. During the ensuing half-century the mathematical imposition of discreteness was progressively elaborated. It is probably assumed by a majority of physicists today that the right tools for any discussion of the explanatory limitation are already to hand in the modern formulations of quantum physics. We shall show at length that this is not the case and that therefore the fundamental problem of atomicity remains. The physical elements of the constructive scheme will need to have a simple mathematical representation that will specify their interactions with one another. One scheme like this has dominated our approach and will be developed in detail. We give it a central place because it provides a theory of the coupling constants and gives the interaction strengths of the fundamental fields; an understanding of the distribution of their numerical values; and in the electromagnetic case a very accurate calculation of its experimental value.

These numbers are dimensionless and therefore there is no contradiction in taking them as the required basis for a discrete scheme for all measurements, so that their primacy is assured. To calculate the numbers it is necessary to postulate a statistical background out of which a hierarchy of nested structures or *levels* develops during the interaction process. The construction of elements of these levels is a two-way process that can be reversed so that it can keep on starting all over again. What dictates what happens in the construction is the interaction with the statistical background, so that the background is the reality whose empirical shape is emerging. However this is the only access we have to the background and it would be wrong to think that discoveries are being made about the distribution of things in a space. The notion of things being "out there" at different distances and in different directions has not yet emerged.

In the 'platonic receptacle' view of space, space is what holds whatever we care to put into it. The idea, particularly of the Timaeus, is that there has to be a 'seat' that things can enter into or disappear from. The 'receptacle' is then the spatial location for them.

In physics a method that we use all the time is to have a variety of arguments of pieces of mathematical analysis each of which is supposed to contribute to our understanding of what is going on at a particular place and yet which are all different and may even be incompatible with each other. All that holds them together is their referring to that point in space. This use of the absolute space seems best put in terms of the platonic receptacle, and in this way the receptacle view is deep in our thinking. As an example suppose we are considering the dynamics of the spiral nebula. It is as though we transport ourselves in the imagination to that place and think what would happen to us in order to formulate the theoretical description. It is now known that the spiral nebulae rotate like solid bodies, so that parts of the nebula at different distances out on a radius stay on that radius. Not merely that, but there may be no simple relation between the rotation and the swept back look: in certain cases the arms may advance point first. After those shocks one is almost disappointed to find that the thing does rotate in the plane in which its arms lie, like a catherine-wheel. Yet, one would have sworn, if ever there were an object whose rough dynamics was plain for all to see, it would have been the spiral nebula – a loose aggregate of matter having some angular momentum, and with some radial motion shown by distribution of material along the arms, but with the arms appearing swept back because of greater angular velocities near the centre.

We are using the fascinating subject of galactic dynamics only to make a methodological point. Our habit is to assume that the universe out there would seem as it does to us here, and exhibit the same mechanics, if we could be transported there. We know we have to give up that presumption as we approach the atomic scale. It is more difficult to give up in the face of possible contrary evidence at the increasingly large scale. It is something between a guiding principle and a prejudice and it should be scrutinized in the light of the information we actually have each time we use it.

Our combinatorial approach makes this scrutiny of the platonic receptacle the more necessary and will figure in later discussions of particle physics.

Chapter 3

The story of the particle concept

Contents

History of the particle concept, Democritus, Aristotle, Epicurus. Reality or hypothetical view? Is there a limit to cutting things up ever smaller? Newton represented the unphilosophical mechanical view that dominated classical mechanics. Dalton confirmed the realism. Planck imposed the discreteness mathematically. Logical incompatibility of discreteness and this classical picture led to the quantum theory of 1926 based on the state/observer principle. Schroedinger's and Dirac's views of this principle. An entirely new attribute of the particle appears with 'exchange'.

The physical world has through modern times been imagined by the physicist as a compendium of moving parts, and these parts move in ways that have a regularity and close similarity that tells him what the world is like and what it can do. This knowledge is enshrined in the laws of classical dynamics. Classical dynamics was formulated from experience of laboratory-scale bodies, in terms of a set of interlinked concepts — most notably mass, distance, momentum, and dimensionality. This set of generative concepts and its use as a universally applicable descriptive language, was the work of a set of thinkers among whom Galileo was the great innovator. Galileo also has something like the modern notion of 'force' but that was not quite the clear modern notion because it had extraneous elements and it waited to be tidied up by Newton. The set of concepts forming this descriptive language constitutes probably the most amazing and powerful achievement of the human mind.

One striking aspect of the set of generative concepts is their evident closed character:- the way that each is linked to others, with the linkages reproducing the same set. For example force requires for its specification mass and acceleration: mass has to be measured through the force it exerts under acceleration: and so on.

So successful was classical dynamics that changes to it and limitations to its applicability seemed unthinkable and this had consequences for the development of the quantum theory that we well know. To this scheme was added the idea of a smallest bit: together with the idea that there were a small number of different kinds of bit: and that the bits of any one kind were identical. These last ideas were assumed in the classical theory though possibly as a hypothesis. After Planck and Thomson and the others, it became accepted experimental knowledge. People would have said that if there were particles they would naturally obey the laws of classical dynamics since all went on in one world.

We arrive in this book at a view of the particle that is a far cry from the view just outlined of it as just a small bit of matter. The new view needs careful statement and in this chapter we first look in general terms at its main features. We start by placing it in its historical context.

Democritus (460–370 B.C.), is the most important figure in the early history of atomicity. He evidently owed much to his teacher Leucippus. Democritus separated space, or the void, from "Being" or the phenomena that went on in the void. The void, for him, was an infinite space in which there moved an infinite number of atoms ("atoma" means *indivisible*). His atoms are eternal and indivisible, and so small that their size cannot be diminished. The idea appears at this stage, therefore, that in cutting things up smaller one reaches a limit. This idea of course appears extensively in later physics, and whether or not Democritus is really the origin of it he does not seem to have analysed it or said more about where it came from. The different qualities of the atoms were meant by Democritus to be explained as different combinations or configuration of the atoms though he seems not to have elaborated on this vital idea but rather to have returned in a circular manner to giving the different kinds special properties in their own right (roughness or smoothness, for example).

Aristotle was critical of Democritus for introducing the idea of 'minima' or entities of the smallest conceivable size. He supported the 4-element theory (earth, air, fire and water) and discussed the transformations of one into another but does not seem to have exploited the analysis of nature into 'atoma' and is therefore hardly a part of the development of the particle concept.

The generality of principles from which Democritus begins is helpful for us and tends to be lost in the subsequent specialization of the principles as the particle began to take its modern forms. We find this narrowing down beginning with Epicurus. Epicurus (*c.* 310 B.C.) believed that everything in the world had a natural explanation, which meant an explanation in materialist terms. The idea seems to have crept in at that stage that explanations in terms of psychically induced, or caused, effects were not to be considered as proper explanations. Evidently 'explanation' was to be material explanation. For Epicurus the universe is finite and eternal and everything in the universe is made up of atoms. Further; even living things are made up of atoms, and it should be possible to explain the mind and its senses through the interaction of these.

In support of Epicurus' development from Democritus, one could see by carefully observing nature, that there were in fact a far larger number of different elements, and by 400 B.C. it was commonly believed that these diverse elements existed as individual particles or *atoms*. A further step would have been to say that the diversity of elements provided a number of different sorts, but that there would be many of any given sort. These would be all the same, or identical. The universe consists of bodies and the void. The bodies may be compound in which case they are composed of elements (atoma) that are indivisible and unchangeable. The elements are not all of the same size, but there is no suggestion that they fall into classes whose members are of the same size.

What can be said in general terms about the views of the ancients that led to the concept of the particle? The most important step in the path toward the modern particle that became current with them was the idea of different classes, each class having a unique type of member but with members of that type all the same. This idea was assumed in the whole of later practice. It derived its force from the view that the process of cutting up the world ever smaller must have a limit. Given that whole point of view, the term 'particle' had become appropriate.

Did the early thinkers believe that particles were real, or might they be hypothetical? We are familiar with one argument in the physics of modern times that holds that the real world is a continuum with fields pre-eminent, and that atomicity is a theoretical imposition that may well be wrong, and we know that that view had a lot of force in the last century. More recently the emphasis has swung back to realism over the particle. Both views were embraced by various of the early thinkers and the tension between the views was always there.

On this matter of the particle, history seems to take a leap to the sixteenth or seventeenth century. Almost everything in mechanics can be traced to the monumental work of Galileo, but Galileo wrote about bodies without bothering very much about limits to their size, and therefore whether they were particles. For that concern we move to his close successor Newton.

With Newton we plunge into the crudest possible material picture of the particle. His very well-known opinion that in the beginning God formed matter in solid, massy, hard, impenetrable moveable particles (see Chapter 4 for full quotation) used the God-language to circumvent any discussion of the deep questions about the particle concept that have occupied us. Particles were little balls with gravitational and inertial mass. His laws demanded matter cut up into bits (as had the mechanics of Galileo) and the space in which the bits were spread out is usually referred to as 'absolute space'. This term is probably best taken to mean that no consideration needed to be provided of the relations between bodies or particles that may be needed to fit them into space. Newton's position was in contrast to that of Leibniz for whom space was relational (see Chapter 2). Newton and Leibniz were contemporaries and engaged in close exchanges of view over the space question as is very well known, but there seems to have been little exchange between the two over particles. Whereas Spinoza saw the world as simple comprehensive substance, and Descartes had extended matter, Leibniz postulated many discrete particles — each being simple, active and independent of every other. These were examples of his monads. The wide sweep of the monadology made no impact on Newton who was content to leave the particle concept unanalysed. Evidently the excitement and exhilarating sweep of the new mechanics (perhaps the most remarkable achievement of human thought) was allowed to carry the conviction that here one had the description of reality direct, so that no one bothered with the doubt that it might not apply universally and on all scales of magnitude — down indeed to the particles.

The contemporary of Newton, Boscovich (1711–1787) took an important step in the history of the particle concept. In contrast with Newton who did not analyse the actual impact of particles on one another, Boscovich, in his "Theory of Natural Philosophy" (1763), thought about the forces that must accompany the impacts in a way that was so modern as to mark a real change. It is exactly as though Boscovich were describing electromagnetic forces, though of course the concepts of electromagnetism did not then exist. He inferred that there must be forces of attraction and repulsion at work between particles of different kinds that had to be specified in detail, and was explicit about the changes from attraction to repulsion that would have to be precisely associated with the varying distances between the particles at each moment, if the particles were to have their mechanical stability as the centres of fields of force.

The centuries following Newton are a history of how sophistication over the particle concept was forced by experimental facts that fell outside the Newtonian picture. The steps are so well known as not to need much comment here, but our (not so hidden) agenda will be to show, emerging outside and beyond the accepted picture, the need for the view of the particle that appears in the rest of the book (starting with the process theory of Chapter 6).

The studies in chemistry by Dalton showed that the atoms had indeed the material reality demanded by Newton, but revitalized older conjectures that we described earlier that matter was constructed out of different species of particle — all the members of one species being the same. It was common for people who held Newton's views nevertheless to hold that atoms were fictional structures in continuum physics, but Dalton's step — taking matter to be constructed out of different species of particle, refuted that position.

Now we take another step. The discovery of the electron made the lingering idea that atoms were a convenient fiction in spite of Dalton, out of the question. Later, Millikan's observation of discrete charges on oil drops, where their motion advanced in jumps would force the idea that they were being impelled by one, or by two, or by more units of charge (later seen to be electrons) reacting to a field. First, however, the guiding spirit was that of Planck. His was the simple approach of forcing on the experimental world a *mathematically* specified discreteness in response to an experimental need and the precision of the agreement of his numbers with experiment forced it to be taken seriously against vigorously expressed objections that no explanation of the mathematics had been provided (*Meine Herren: dass ist kein Physik!*). The difficulty in jumping from any numerical scheme to physics was discussed in the last chapter.

The problems posed by Planck's work persisted up to the Bohr atom: Bohr put some classical dynamics into the behaviour of the particles composing the atoms. The theory was by then so successful that it had in some way to be right in spite of its incomprehensibility. To get rid of this incomprehensibility was the task of the quantum theory. The success or failure of the quantum theory in this task will be assessed in Chapter 11 where we address what we call the state/observer philosophy — taking that to be its central argument. We shall find this philosophy to fail — at any rate so far as the understanding of the discreteness of the elementary particle is concerned.

Our present purpose is only to pick out from current thinking on quantum physics those changes that it has made in the concept of the particle itself. There are some such changes whose evident importance is not acknowledged. Thus in quantum mechanics the Schrödinger equation is usually taught as describing the behaviour of *one* particle. However when the theory is "made relativistic" in the form of Dirac's equation, it turns out to be impossible to look at it that way (notwithstanding that it is usually called "Dirac's equation for the electron").

The most important change follows from the appearance of "exchange". Exchange will be discussed in Chapter 8 in the rôle it plays in the theory of fundamental particles and in particular in the quantum theory. For example the concept of exchange provides a background into which one can fit the modern discovery of the mesons — the scope of the concept being expanded as more mesons were found.

That is the conventional view of exchange. However we wish to show that the very appearance of exchange altered the idea of the particle itself, and did this in a way that our theory made comprehensible. We started with the identification of our basic level multiplicities as the origin of the coupling constants. The fine-structure constant is the most prominent of the coupling constants. It is usually presented as a dimensionless ratio of atomic constants that are individually measures of particle properties. We calculate for it a very exact value, but that calculation is quite independent of those particle measures, and they are only associated with it later on. That constant does indeed come from a stage in the developing construction of every particle but the existence of each such particle depends on there being more than one of them in combination. We could therefore say that the exchange concept is built in. It marks the point at which the particle idea swings in the direction of the process theory. With exchange, particles cease to be the unanalysable individuals that are reached at the end of a variety of experimental investigations, and take on properties that derive from the mere existence of two at the same time. These properties are thus mutual. As described in Chapter 8, a mutual force is explained away by postulating a third particle that has the new rôle of travelling back and forth between the original two. An energy goes along with this motion and since there is energy there will be force. No details are given of the mechanics of this travelling back and forth motion, and this is because it is all just a *façon de parler*. It is much clearer to explain the exchange idea as something quite new.

The fact is that the particle concept as it existed during the reign of quantum theory is not adequate to handle exchange, and we ask whether our particle picture can do any better. Our answer is 'yes' because the essentials of the exchange concept exist through the hierarchical structure. In that, the quadratic group representing the primary dimensional structure is repeated at a second level. Then the combined group structure can be used as a unit, or else we can persist in using both separately. The process view requires us to combine both views — switching between them. There lies the exchange notion: you need both, but simultaneously have to treat them singly. Indeed this picture does not provide the particle buzzing back and forth, but as we have argued something would be wrong if it did.

The appearance of recondite kinds of particle properties brings to mind the quark. This is indeed needed to complete the current picture of atomic constitution but the quark cannot be said to be a part of the *spatial* particle picture. One may insist that with quarks the particle picture has had to change, but since they cannot be isolated they are something different and we leave discussion of them and their conceptual place to later chapters.

Let us anticipate the conclusions of this book to see where its argument takes the particle concept. A fundamental construction process generates levels of increasing size. The evidence is strong that the ordinal numbers that give the magnitudes of these levels correspond with the characteristic numbers (coupling constants) of fields or particles. Could it be right to say that the elements of the levels are themselves particles? The answer has to be NO if by 'particle' we mean a free-standing entity capable of moving about among others. The generative interactions are mathematical entities that have only some of the properties implied by the word 'particle'. In particular there is no space at the stage where the experimental connections with physics are derived. How this limited account is extended to provide for ordinary dynamics will be explained in Chapter 9. It is important to realize however that the first incomplete stage is inevitable in the provision of the particles, and that some of their essential properties have already appeared at that stage.

The discovery of the constants and the great part played by Parker-Rhodes was indeed prompted by physical theory rather than through the development of the process theory as we have portrayed the history in this chapter. That appears in Chapters 4 and 5.

The elementary interactions in the 'process' thus have to have a sort

of physical reality even though we deny them the status of fully-fledged
particles precisely because there is no way of getting to fully-fledged particles except by passing through that intermediate stage. This progression
in stages to the particle is without parallel in current physics and it is the
conclusion to our story of the particle concept.

Chapter 4

Dimensionality

Contents

Eddington's program of finding values for dimensionless constants. Symmetry of spatial dimensions invoked in answer. The quadratic group S fundamental. Complexity had to be introduced by putting together a multiplicity of groups. Association of two, gave enough complexity to define electromagnetism. No metric used. Multiple feedback models constructed (Gordon Pask) to provide a metric. Parker-Rhodes formalized these principles and discovered the algebraic hierarchy with its striking progression of numbers.

The last chapter was a historical survey of the development of the particle concept. The present, by contrast, concentrates on the way the authors have developed the combinatorial ideas, and will be expanded later.

The thinking of Eddington played an important part in the discussions from which this book arose, but the theory expounded here is totally different from Eddington's. Apart from his important work in introducing and explaining Einstein, Eddington was well-known for his view that much that was accepted as theoretical analysis of experiment was really imposed by the way we thought about it. In contrast with us, Eddington defined 'particle' as "a conceptual carrier of a set of variates". The word 'variates' probably is just a hangover from Eddington's use of statistical methods to analyse astronomical observation. 'Conceptual' probably is his way of saying that he is introducing his philosophical position. Eddington called his position "selective subjectivism". This position was supported, in Eddington's view, by the possibility of obtaining certain important constants prior to experiment. Not surprisingly, this view drew upon itself criticism amounting to scorn for seeming to replace proper empirical method with what was labelled "a priori" thinking. In fact 'a priori' became a convenient word to dismiss Eddington. We shall find, in this chapter, reasons to reconsider Eddington's arguments: the basic numbers that we find generating particles are reminiscent of his position.

In the period soon after Eddington's death, the authors were persuaded of the correctness of his policy of obtaining numbers characteristic of foundational physical theory from algebraic considerations. His efforts in "The Relativity Theory of Protons and Electrons" and later in "Fundamental Theory" could not be sustained. Nevertheless we thought his motivation was well-based — and we tried to find out where the thinking had gone wrong. Eddington's calculations had started from the 3 and the 3 + 1 of space and space-time as necessitated by the structure of a Clifford algebra. Accordingly he saw them as existing in their own right beyond their geometrical setting. This procedure seemed to us to require an investigation of what dimensionality itself really was.

Somehow some vital numbers are introduced into physics at the inception of efforts to explain any properties of particles. Famously, Newton[7] said "All these things being consider'd, it seems probable to me, that God in the Beginning form'd matter in solid, massy, hard, impenetrable, moveable Particles, of such Sizes and Figures, and with such other Properties, and in such Proportion to space, as most conduced to the End for which he (sic) formed them; and that these primitive Particles being Solids, are incomparably harder than any porous Bodies compounded of them; even so very hard, as never to wear or break in pieces; no ordinary Power being able to divide what God himself made one in the first Creation". Modern physicists would look for the continuing identity of particles of a given sort to be understood within the ambit of quantum theory, but they follow Newton in accepting the sorts of particles as a given fact of Nature (or God).

Our seminal idea came as follows. We said that when we described the world as spatially three-dimensional we were simply drawing attention to a very strong symmetry. If we say that there are a number, j, of dimensions then we assert that if we build a set of relationships from these, then each such relationship remains true in whatever order we take the j dimensions. This symmetry was best shown at its simplest in an algebraic structure with a single binary operation. The significance of this binary operation was not yet apparent. The symmetry then requires that any equation $pq = r$ which holds in the structure will imply that $qp = r$ so that the operation is commutative. But from $pq = qp$ it will follow that $p^2 = q^2$ so that all the elements have the same square; call it e. Moreover $p^2 = e$ implies that $pe = ep = p$, so that e is a (two-sided) identity. Finally, if $p(qr) = s$, then $r(qp) = s$ and commutativity leads to associativity, $p(qr) = (pq)r$. We are therefore dealing with an Abelian group in which every element is of order two. It is well-known that such a group is a direct product of cyclic groups of order two: $C_2 \times C_2 \times \ldots \times C_2$. C_2 is itself a trivial example of the symmetry so that the simplest non-trivial structure is $C_2 \times C_2$ which is the quadratic group (Klein's 'Viergruppe') S.

However the three elements a, b, c also possess a stronger symmetry. If any equation $pq = r$ holds in S, it remains true if any of a, b, c in the equation are replaced by other elements (as long as different elements are replaced by different elements). Thus $a^2 = e$ implies $b^2 = e$. S is unique in having this symmetry. Here, then, are the primary numbers 3 and 4.

The question that then confronted us was how 'could such a view of dimensionality replace the usual way of representing the results of measurements as ordered sets — namely by coordinates?' Such sets could not be based upon the a, b, c and e of the group S. It appeared that the only way forward in increasing the range of numbers available to represent measurement was to associate two S-structures to form a product algebraic language, so that one could, as it were, change gear and go up a stage of complexity. The idea of hierarchical structure was born.

What was still missing in our thinking at this stage was how exactly to relate two structures at different levels. We simply supposed that the complexification process (attaining greater descriptive power by combining as many of the structures already created) could be continued. The treatment of a coordinate space remained for future thinking.

The relationship between mechanics and electromagnetism can satisfactorily be understood in terms of this progression of stages of complexity, as will be shown in Chapter 7. We called this primary representation abstracted from any coordinate space a *theory-language*, as we have described in Chapter 2. Eddington could be said to have used the level idea in his calculation of the fine-structure constant since he used the numerical dimensions of space-time in juxtaposition with the 16-term energy-momentum tensor. He did not use the two as stages in a build-up process as we find essential however. Not only did he not see how to relate the levels (and at this point neither did we) but neither could he explain how the hierarchical process of going from 16 to 256 elements, and in one calculation to $256^2 = 65536$, would ever terminate. He was reduced to the rather weak statement that physics does not seem to need a greater level of complexity. We refer to this as 'Eddington's difficulty' and return to it in the next chapter. To our way of thinking it was essential that each new level should be constructed out of the operations or operators that described what was going on in the previous level, so that nothing new and inevitably arbitrary should be brought in, though we could not then see how to do that.

The next question was how we should incorporate the empirical input to the structure. How could the numbers that gave the results of measurements find a place? There were evidently 4 interacting elements at the first level and 16 at the next, and we knew absolutely nothing about their structure except that three of those at the first level were equivalent. We thought a model might be constructed that would represent the way the elements of the levels continuously interacted with each other, while rigorously preserving the basic symmetries. Then we might impose constraints on the activity of the model to introduce a 3-and-one asymmetry artificially. Then we might find that that restraint was in some way responsible for the values of the dimensionless coupling constants that we had in our sights, and that the results of other measurements that had not the total fixity of being dimensionless might follow a similar pattern.

Gordon Pask was interested in the idea. He thought that one could build a machine in which (a) every element was exactly the same as every other, and (b) each could be connected with every other in such a way as to run through all the possibilities of connexion. Pask provided $4 + 16$ units which were high-grade amplifying switching units that had been sold off as unneeded wartime equipment. The feedback loops connected the 16 units with each other in pairs *via* the 4 units, and the connections were discrete — being controlled by switches that brought in the discreteness. The choice of connections that existed at any given time was controlled by the state of the units being connected. Pask thought that if you connected everything with everything else while observing their total symmetry you were representing everything that could happen and therefore introducing nothing arbitrary.

These units, using thermionic valves of course, were quite large, and the whole lot together with the ancillary switching units sat on benches occupying a large room. They used quite a lot of power — several kilowatts. This wondrous machine continually changed its configuration and the activity could be heard going from place to place in it. This technique appeared in most of Pask's actual learning and teaching machines. Input to the 16-level was randomised to compensate for having to cut off the construction somewhere.

It was intended to use the machine by imposing a constraint on its working corresponding to the physical requirement that one of the 4 units differed from the others — distinguishing space from time. One would then see if this imposition produced numerical effects that might be interpreted as the coupling constants, and perhaps exhibit these in a larger class of numbers that were more like continuous variables.

Bastin had been offered the use of the PACE analogue computing equipment in Brussels in the hope that that would make our investigation more precise. At this stage Frederick Parker-Rhodes came on the scene and Parker-Rhodes accompanied Bastin on one trip. He was to come again but was prevented by flu. He had already maintained that we seemed to be doing what binary algebra was designed for. He used strings of binary units, written 1, and 0, and used them to express the ideas of a hierarchy of levels and of the construction of each given level out of elements of the preceding one that was advocated by the authors.

When Bastin got back Parker-Rhodes had worked out the essentials of the bit-string hierarchy which is described in the next chapter. It was noticed at once (including by Parker-Rhodes) that the numbers that characterized each stage in the generation of levels were very suggestive of coupling constants of particles with their fields. The story has got about that Bastin asked Parker-Rhodes to produce numbers like these and that he succeeded in doing that by fiddling about with algebra. This tale does not do justice — particularly to Parker-Rhodes. In fact he was explicitly trying to represent in binary algebra the two principles:- (1) hierarchy construction in levels of increasing complexity, and (2) the construction of each new level using as elements the operations at the previous level.

Another story is entirely accurate. A few days after this denoument Parker-Rhodes came, deflated and actually apologetic, because he had come upon the fact that his hierarchy progression stopped. We commented that he had discovered why gravitation is the weakest force and was seeing that as failure!

No more was done with the mechanical model since all energies were turned for some time to exploring the consequences of Parker-Rhodes' algebra.

Chapter 5

The simple bit-string picture

Contents

A hierarchy of levels used the idea of any one level being constructed out of the operations at the preceding level. Parker-Rhodes gave this idea digital or combinatorial form. The unit for such physics was the discriminately closed subset. Characteristic numbers emerged. These provided the setting for particles.

This chapter describes the early history of the idea of the *levels* that came to assume such importance in the process theory that principally occupies this book. One would have to say that current physics is a *one-level* theory. Of course if there is only one level then the idea of level hardly comes into play. One might perhaps imagine that electromagnetism had been thought of as a new and different kind of thing altogether from classical mechanics. Then the 'forces' between charges might have had to be built up quite differently from the forces between masses. One word 'force' would no longer suffice. Physicists do not think this way since the unification of these forces is regarded as a great success.

A quite different way of thinking would appear if one had always to construct a piece of theory of limited application as a first stage, and only then to build upon and use what one had done in order to encompass a wider range of experimental knowledge. One would then have more than one level and the resulting system could be described as *hierarchical*. This different way of thinking has been employed in the process approach of this book, and we look now at the way it had already appeared before a concerted process view had been formulated which later came to incorporate it. The writers did have such a theoretical nucleus given by their studies of the symmetry properties of S, the quadratic group, and it seemed to them that the only way to combine such structures was by appealing to a system of levels.

At this point Frederick Parker-Rhodes, in the context explained in the last chapter, took up the idea of basing physical description on a hierarchy of levels in which a new level could be constructed using operators at the existing level to form new elements of greater complexity at a new level. (This construction was to be the way increasingly large numbers representing physical quantities could come to exist. Later, it became clear that these levels should be interpreted as the set of particle interaction strengths, or of the fundamental fields.) Parker-Rhodes formalised the ideas. He introduced for the first time a completely discrete or combinatorial calculus in which things in the world were represented by strings or vectors of symbols 0, 1 (bit-strings). Any two vectors of a given length could be brought together to produce a new one by operating with the two elements in a given place using the symmetric difference operation, that is, addition over the field Z_2 with only two elements. A new vector would arise if the starting vectors were different; otherwise a string of zeroes would be produced indicating that they were the same. This process later came to be called *discrimination*.

Physical importance would attach to any set of vectors that was closed under the discrimination operation applied to any two different vectors. This was called a *discriminately closed set*. Every set of vectors will generate its *discriminate closure* by repeated application of discrimination to every two different vectors. In the world this will happen given enough time.

So far, Parker-Rhodes' structure is just an isomorphic copy of the $C_2 \times C_2 \times \ldots \times C_2$ groups found by the authors. He represents the (Abelian) group representation by addition over Z_2 and accordingly e becomes the zero, the identity element in addition. What was important in Parker-Rhodes' formulation was that it provided the precise relationship between levels. Moreover it turned out to have solved Eddington's difficulty over indefinite continuation because Parker-Rhodes' sequence terminated.

Operations upon the vectors were defined as square matrices over the binary field — thus introducing a second operation written as multiplication. A discriminately closed subset generated by a subset of the generators of a level was represented by a square matrix leaving every element of the subset unchanged. These matrices were therefore the generators of the new level and for that purpose they were rewritten as column or row vectors of squared order. Elements of the new level thus have to be mapped onto a space of order of the square of the preceding level. Since the number of significant entities increases faster than the size of the vector space the process comes to an end.

Because physical importance will be attached to any set of vectors that is closed under the discrimination operation there must be a limit to the number of discriminately closed sets at any given level, and this limit is the characteristic number of that level. Parker-Rhodes drew our attention to this sequence of these numbers, which has the value 3 in the simplest case and thereafter follows the rule $2^r - 1$, where r is the number computed for the previous level. If we take cumulative sums of the numbers at the levels to give the total available numbers then these come out $3, 10, 137,$ and $\approx 10^{39}$. Parker-Rhodes could not justify the formation of these cumulative sums, but as we shall see it was an intimation of our fundamental principle of *process*.

The approximate correspondence of these numbers with the reciprocals of the coupling constants of the fundamental fields, including the astonishing gap between the electromagnetic and the gravitational numbers, and with the fact that the sequence cannot be continued, has been sufficiently persuasive to promote further work over a long period.

Parker-Rhodes originally saw the discriminate closure of the sets by analogy with the eigenvectors of quantum theory, but we have preferred the terminology that was to come later. In particular the idea of the vectors being physical things that reacted to others in an unknown statistical background was foreign to Parker-Rhodes' thinking. Parker-Rhodes had succeeded in constructing a system with levels and solved the level-change principle that was necessary to incorporate the basic principles used by the authors in a way that placed it within conventional theory. Of course it raised as many problems as it answered. There were no clear physical answers to the obvious queries:- why bit strings? Why start with just two bit strings? Why square matrices for level change? Why eigenvectors? Why form cumulative sums of 3, 7, 127?

It will be shown in the next chapter that the process theory yields a more complex algebraic structure. This structure contains the Parker-Rhodes algebra in a peculiar way. By exhibiting the Parker-Rhodes algebra in this way we answer the questions above. We present it here in spite of its crudity so that readers whose interest is primarily in the physics will be able to see what the whole argument is about.

The setting in physical theory of the numbers is given in Chapter 12, while the full calculations of the most important and familiar of the numbers appear in Chapter 6. It may seem strange that so much can be said about the precise values of the constants before their place in current physical theory is described. This apparent presentational confusion is actually what is demanded by our general approach. We do in fact start from algebraically generated numbers which will provide our mode of construction of the physical world through their progressive development, and that development is the world of physical theory.

Chapter 6

The fine-structure constant calculated

Contents

The inverse fine-structure constant is calculated as $137.036011393\ldots$, *agreeing to better than one part in* 10^7 *with the latest experimental value,* $137.035999710\ldots$. *The statistics that generates the base number* 137 *from the process theory is progressiviely refined in a long calculation, the details of which are set out at length in two notes at the end.*

In earlier chapters we described the historical path that led us to a progressive construction that had stages in which we found the numerical characteristics of fundamental particles. This constructive picture is unfamiliar in physical theory and if we wish to maintain the numerical successes obtained this way we have to explain the constructive picture. The present chapter represents a change of gear. In it we ignore the historical development in order to provide a formal basis for the construction.

The chapter falls into two closely related parts. The first part fulfills the promise of the last chapter by exhibiting the Parker-Rhodes algebra as determined in a particular way by the more general algebraic structure derived from the notion of process. The second part then calculates a close approximation to the fine-structure constant. In this calculation an important part is played by the two related algebras, that derived from process, and Parker-Rhodes' original one.

We postulate a universe that is constructed progressively, and so we speak of process. We expect physical principles to emerge from our formulation of the process. These principles will be valid for all experimental procedures and will guide them. In the process elements are constructed in a sequence. It is convenient to speak of 'elements coming into play'. To start with, we do not know what meaning to give them: any meaning that they will eventually acquire is solely as a consequence of the construction process and the formalism develops with the construction. This approach is reminiscent of Brouwer, and Brouwer has been an influence on the authors though this theory is not intuitionistic. To be a sequence implies that the process determines whether a new element is the same as one already in play or not. This determination is of paramount importance in our theory. Such an emphasis is by no means novel. One finds it already in the nineteenth century with Sir William Hamilton. John Stuart Mill summarises Hamilton's view with approval: "We only know anything by knowing it as distinguished from something else; all consciousness is of difference; two objects are the smallest number required to constitute consciousness; a thing is only seen to be what it is by contrast with what it is not." (in his *Examination of Sir William Hamilton's Philosophy*). The difference from our process theory is that we have abstracted from the epistemic character of Hamilton's statement.

However the determination is done, finding that b is the same or different from a is a new element, c say, in play. The binary operation $a\,b = c$ is called *discrimination*, and a primary object of our investigation is to determine what this operation is. It will turn out that it has to have different forms at different stages in the theory. When discrimination has been formalised it cannot be restricted to apply to discriminating between new and old elements only (since to do this would require the process to determine whether or not one of the two elements was indeed new or not and so an infinite regress would result). The discrimination must go on all the time so the process must be iterative. In the course of the construction elements may be deleted as well as coming into play.

The process continues without limit so there is a sense in which one might say it was potentially infinite, although we shall find that this continual development is restricted in a particular way. This is the Parker-Rhodes limit (see below). Here, a nod in the direction of Brouwer again since the finiteness is such that for each element to come into play the determination of its place in the sequence takes only a finite number of steps in the process.

A binary operation suggests a group in which the binary operation is the discrimination operation. But this will be an example only if the group satisfies the condition that the set of group elements that can arise as products of 2 different elements is disjoint from the set of products of equal elements (i.e. squares). Now think of the Cayley table (multiplication table) of the group. The restriction is that the diagonal elements can never occur in the rest of the table. But since each line of the table has all the group elements in it this means that there can be only one element occurring in the diagonal which must therefore be the identity. Hence every element is of order 2 and it is well known that the group is a direct product of cyclic' groups of order 2. The identity is a signal that the product is of equal elements.

Another way of writing such a group is as a set of bit-strings (vectors over the field Z_2). The signal for equality is now the zero string. It is not necessary for the system to be a group. Another example would be to begin with the quaternion group in the abbreviated form familiar from quaternion algebra. This has three elements and an identity with their four negatives. Now modify the Cayley table by writing the identity in all diagonal places. Then $a(-a) = 1$ in the original table becomes $a(-a) = -a^2 = -1$ in the new one. The identity would then be the signal for equality, and this fulfils the condition above because the identity never appears elsewhere in the Cayley table. Just as the group had an isomorphic copy made of bit-strings so here there is a bit-string version with strings of length 3. One way of doing this is to number off the bit strings like this:

$$100 \quad 010 \quad 110 \quad 001 \quad 101 \quad 011 \quad 111$$
$$1 \quad\;\; 2 \quad\;\; 3 \quad\;\; 4 \quad\;\; 5 \quad\;\; 6 \quad\;\; 7$$

that is, in binary, though in the reverse order from the more usual one, and then to define $a \cdot b = a + b + H(a, b)$ where H is called the *complexion* of the pair (a, b) and is defined by

$$H(1, 2) = H(2, 3) = H(3, 1) = H(u, u) = 0$$
$$H(u, v) + H(v, u) = 7, \quad \text{and} \quad H(u + 7, v) = H(u, v)$$

The system is not a group because it is not associative. It is a loop. We call it Q^*.

Because the argument leading to the complete determination of the discrimination operation is lengthy we begin with a summary of it, with the main conclusions. This will also introduce some technical terms. In the flux of discriminations a set of elements will usually be augmented or diminished. Physical importance attaches to sets that are not increased by random discriminations between the elements :

A *discriminately closed subset* (dcs) is a set of elements closed under discrimination of any two different elements of the set. This definition needs further qualification when the subset has one or two elements only. Details must be deferred to the full discussion.

The *discriminate closure* of any set of elements is the smallest dcs containing them.

A set of *generators* of a dcs is any set for which it is the discriminate closure.

A level of type r, L_r say, is a set G of r elements, called *generators* of the level; together with the members of the level which are

    (a)   the elements of the discriminate closure of G,

    (b)   all those dcss which are discriminate closures of subsets of G (including G itself).

The *multiplicity* of L_r is the number of different dcss which are members of L_r.

When all members of L_r have come into play the process has *filled* L_r.

If further elements come into play the level is inadequate to describe the further ranges of experiment and an increase in structure is required. Such an increase is provided by :

The *superior* level to L_r is that defined by a set of generators, one for each of the dcss of L_r.

For this increase of structure the process *goes up a level* to the superior level to L_r. The elements of this superior level are the dcss of the original level (together with their dcss of course).

The general picture will now be clear. The process starts with some level, which becomes filled eventually. It then goes up a level and repeats the filling up with the new elements (the dcss of the original ones). The detailed investigation shows that the construction of new levels stops quite soon for technical reasons that are of great physical importance. In one case only, when the initial level is one with two generators, a four-fold hierarchy is built up. In Parker-Rhodes' original construction the successive multiplicities of the whole system were 3, 10, 137, and $10^{38.2}$. The approximate relation of at least the last two to the reciprocals of the interaction strengths is important for the connection to physics. We shall find analogous results in the exact theory.

In the case of a group a dcs is simply a subgroup with the identity omitted. If the group is, for example, of order 8 — $C_2 \times C_2 \times C_2$ — a typical dcs is $C_2 \times C_2$ of order 4 except that the identity is to be omitted. That is, it is the non-identity part of the quadratic group $[a, b, c]$ where $a b = c$ and so on.

In the bit-string expression with strings of length n a typical dcs is the set of bit-strings of length $m < n$. Notice however that in each example the dcs may not be presented in the way just described. For example if $n = 3$ one dcs would be 010, 101, 111 which does not appear as a bit-string of length 2. But it is equivalent to it since the first and third bits are the same, so that one could be omitted. Similar qualifications apply to the group.

The dcss in the loop case are left to the detailed development as are details of what form the superior level has in each case. Suffice it to say here that the elements of the superior level in the bit-string case will be square matrices over Z_2.

We begin the investigation of the form of the discrimination operation by asking first how the process determines whether two elements at the same level are the same or not. If a, b are in fact the same element then no new element can arise in the discrimination. Hence either $a \cdot a = a$ or $a \cdot a$ is not an element. The first alternative cannot achieve discrimination because the next step would have to be to check that the element on the right-hand side was or was not a: that is, exactly the same problem again. The second alternative is the only possible one and $a \cdot a$ is then a non-element, z say, which must be unique so that its appearance removes the need for any further discrimination. We shall prefer the term *signal* to 'non-element' for z.

In order that z may signal equality the binary operation must be such that when a, b are different then $a \cdot b \neq z$. A structure with such a signal is a *discrimination system*. When a, b are different elements their discrimination is then a third element: call it c. Then $c = a \cdot b$ is the discrimination of a putative new element b against one already in play, a. This is different from the discrimination of a putative new element a against b already in play; so $b \cdot a = d \neq c$. However for obvious physical reasons c and d are closely related. We call this relation *duality*; say c, d are *duals*; and write $d = c^*$, $c = d^*$. The notion of duality plays an important part in determining the form of the discrimination operation. It leads to the following further notions: two elements may be (i) the *same* or *identical*, (ii) *duals*, (iii) *essentially different*. Duality has been defined for c and c^* which come from two related discriminations, but the structure of the system is such that every element can be expressed (in many ways) as a discrimination, so that a^* and b^* also exist.

To determine when two elements are duals will require a signal for exactly the same reasons as for when they are identical, and this signal must be different from that indicating identity. We call it y, so that $a \cdot a^* = y$.

The second signal necessitates a change in the definition of a dcs. A dcs is a set of elements, so signals are excluded. The definition should therefore read *any two essentially different elements of the set.*

Duality has certain obvious properties: suppose that $a \cdot b = c$ and consider $a \cdot b^*$. This cannot be c since that is $a \cdot b$, but there are no elements in play other than a, b, c and their duals, so it must be the case that $a \cdot b^* = c^*$. By a similar argument $a^* \cdot b = c^*$ from which follows (taking the two results together) that $a^* \cdot b^* = c$.

It is now clear how to determine the form for discrimination in case (iii) when a, b are essentially different. Suppose $a \cdot b = c$ and consider what value the process must give to $b \cdot c$. Here again the only elements involved are a, b, c and their duals, so it must be the case that $b \cdot c = a$ or $b \cdot c = a^*$. But this second possibility is ruled out because if $b \cdot c = a^*$ follows from $a \cdot b = c$, then $c \cdot a^* = b^*$ will follow from $b \cdot c = a$, and applying this again yields $a^* \cdot b^* = c^*$ which is in conflict with the result above that $a^* \cdot b^* = c$. Hence from $a \cdot b = c$ it follows that $b \cdot c = a$ and therefore that $c \cdot a = b$.

Another way of putting this is that the binary operation $a \cdot b = c$ can be expressed as a triadic relation $R(a, b, c)$ with the property that if R holds for any three elements it also holds for any cyclic permutation of them. These results determine the following table in the way indicated below it.

	a	b	c	c^*	b^*	a^*
a	z	c	b^*	.	.	y
b	c^*	z	a	.	y	.
c	b	a^*	z	y	.	.
c^*	.	.	y	.	.	.
b^*	.	y	.	.	.	.
a^*	y	.	.	.	.	.

(i) The diagonal z's and the anti-diagonal y's follow from definition.

(ii) $a \cdot b$ defines c and so $b \cdot a = c^*$.

(iii) From the symmetry above $b \cdot c = a$, $c \cdot a = b$ and duals.

The rest of the table has not been filled in because it follows at once from $u \cdot v^* = u^* \cdot v = (u \cdot v)^*$. It is clear that this structure, L_2, is Q^*, described above.

The table holds for any two elements a, b. If there were more generators than two, and so a larger table, any two generators would generate a subsystem with the same table. The table is that of a level L_2 of type two, the set G being $[a, b]$, and the members of L_2 being

(i) a, b, $c = ab$, a^*, b^*, and c^*

(ii) The dcss $[a]$, $[b]$, and $[ALL]$

(the third dcs being all the six elements in (i)). The multiplicity is the full $3 = 2^2 - 1$ permitted.

The superior level to L_2, L_3 is generated by discriminating the three dcss of L_2. This provides the necessary increase in complexity of structure. To discover how the discrimination is formalized in this case, we consider first how the process might continue in L_2. If L_2 is full, and a putative new element u comes into play the process cannot determine whether u is in the set of elements of L_2 by discriminating u with each element of the set in turn. For in each successive discrimination it would first be necessary to determine whether the comparison element was one that had already been used. This would lead to an infinite regress. Instead u must be discriminated against the dcs, T say, as a whole. The triadic relation is now $R(T, u, v)$. To determine the form of this consider first the case when u is in fact a member of T. No new element can come into play so either $R(T, u, u)$ or else a new signal must be introduced. In this case there is no need to reject the first possibility since the process has already incorporated discrimination between individual elements and so no infinite regress is involved. We can again put R into the form of an operation and use T unambiguously to denote both the dcs and its operator. If u is in T, then $T u = u$ and correspondingly if u is not in T, then $T u = w$ where w is a different element from u.

There is a further restriction on T arising from the physical importance of dcss. The results of discriminating the elements of a dcs against T must again be a dcs. If u, v are discriminated against T to give w, x, then when $u \cdot v$ is discriminated against T the result must be $w \cdot x$; that is $T(u \cdot v) = Tu \cdot Tv$ so T preserves the original discrimination. Moreover if $u \neq v$ then $Tu \neq Tv$ for, if not, $T(u \cdot v)$ would be the signal z, contrary to $T u$ being an element. So T is an automorphism of the discrimination system. Hence T preserves duality, so that $T u^* = (T u)^*$. This leads to the qualification about dcss mentioned earlier when they have only one or two members. If $T = [a]$, so that $T a = a$, it follows that $T a^* = (T a)^* = a^*$ so that, from the definition of T, a^* is also a member. The process must deal with the single-element dcs in the form $[a, a^*]$, not a alone. This is consistent with the modified definition because $[a, a^*]$ does not contain two essentially different elements.

It is straightforward to prove that for any dcs T there is an automorphism characterising it, in the sense that $T u = u$ if and only if u is in T. We defer the proof till later when it can be carried out more easily. The increase in complexity of structure required by level change is provided by the discrimination between such automorphisms. This discrimination must be consistent with that between the original elements. To set out how this happens note that the two automorphisms T, U will have to be the same if and only if $Tp = Up$ for all elements p in play. The discrimination between T, U, which we write as TU to be consistent with the original discrimination, must be such that $(TU)p = z$ if and only if $T = U$, that is $TpUp = z$ for all p in play. So the discrimination between T, U must be that familiar in mathematics as the induced operation :

$$(TU)p = TpUp, \qquad \text{for all } p.$$

When the process is discrimination between dcss in this way it has *moved up a level*. With this definition of discrimination between automorphisms the new level is again a discrimination system. For this property is defined at the lower level by the condition that, if there exist elements u, v, w such that $u \cdot v = w^2$, then $u = v$. For automorphisms U, V, W such that $UV = W^2$ it follows that $UpVp = WpWp$ and so from the condition at the lower level $Up = Vp$ for all p, which is the definition of $U = V$. We defer the details of the structure of L_3 until later because we want to explain a simplified way of investigating it.

This simplified way is not part of the process but is closely connected with it. Consider the two dcss $[a, a^*]$, $[b, b^*]$ at level L_2. Define a discrimination operation between dcss, not by the "extra structure" discrimination investigated above that the process uses, but by

$$[u, v, w, \ldots] \cdot [u', v', w', \ldots] = [u \cdot u', u \cdot v', v \cdot u', v \cdot v', \ldots].$$

In the case of $[a, a^*]$, $[b, b^*]$ it is clear that

$$[a, a^*].[b, b^*] = [a \cdot b, a \cdot b^*, a^* \cdot b, a^* \cdot b^*] = [c, c^*].$$

Also $[a, a^*].[a, a^*] = [z, y]$
and $[z, y].[a, a^*] = [a, a^*].$

The three elements $[a, a^*]$, $[b, b^*]$, $[c, c^*]$ have the structure of the quadratic group S and $[z, y]$ is the identity. We call this the *characteristic group* of level 2. Its usefulness is that it gives guidance about the way the algebra is going, rather as homology groups do in algebraic topology. It forms the basis of a hierarchy of a simpler kind. Call this lowest level L'_2; its multiplicity is 3, equal to that of L_2. The next level in this simplest hierarchy, L'_3 is then the characteristic group of L_3; the multiplicity of L'_3 and L_3 are both 7, and so on up the hierarchy.

We shall justify this result below when we have established a convenient notation. We call this simpler hierarchy the *skeleton* of the actual one. Because of the close fit between the two hierarchies and because the skeleton is based on dcss, the physical results (numbers) calculated from it are good approximations to the correct ones.

Instead of the quadratic group in abstract form it is more convenient to use the isomorphic copy mentioned above, of bit-strings — vectors over the field Z_2. For the quadratic group the bit-strings are of length 2 and the discrimination operation between them is commutative, being addition of vectors over Z_2. The signal for equality is zero and no signal is needed for duality. The generators for L'_2 are column vectors, more conveniently written as their transposes $(1, 0)$, $(0, 1)$, and discriminating them gives $(1, 1)$. The automorphisms that generate L'_3 are then non-singular 2×2 matrices. The (uniquely determined) one for the dcs $[(1, 0)]$ is the 2×2 matrix transposed as $[(1, 0), (1, 1)]$ and that for $[(0, 1)]$ is $[(1, 1), (0, 1)]$ whilst that for all three elements is the identity $[(1, 0), (0, 1)]$. At the next level, L'_3, it is obvious that the induced discrimination operation defined between the automorphisms is just matrix addition (over Z_2). Then these matrices can be coded as bit-strings of length 4, and so the construction continues with 4×4 matrices and so on. What has been shown here, then, is that the lower levels of the skeleton hierarchy are just those of the original Parker-Rhodes construction. Here "lower levels" refers to the fact that the algebraic structures will agree all the way up the hierarchy, but the approach differs from Parker-Rhodes for reasons that will be explained below.

This result therefore gives the process theory justification for the Parker-Rhodes hierarchy construction, albeit as a close approximation to the actual hierarchy. In the informal account of Chapter 5 the use of bit-strings was motivated by the description of elementary particles in terms of a set of quantum numbers. Here the situation is reversed: the process account defines a non-associative algebra. This algebra has a close approximation with a representation in terms of bit-strings forming the Parker-Rhodes hierarchy. (Here 'representation' is meant in the sense of an isomorphism, not in the sense used in group representation theory.) The idea of a set of quantum numbers is now a derived notion. This derivation of the Parker-Rhodes hierarchy answers the queries about it listed at the end of Chapter 5.

We now proceed to the calculation of the fine-structure constant. It proves convenient to do this first for the Parker-Rhodes algebra, which we have called the skeleton here. Then the correct result can be derived more easily. The small numerical difference between the two results can be taken as an indication that the skeleton is a very close approximation to the true hierarchy.

Detailed calculations in the two hierarchies require a better notation.

This is provided for the skeleton by Conway's nym sum. We write 1, 2, 3 for the bit-strings $(1,0)$, $(0,1)$, $(1,1)$, the connexion in general being that a bit-string a, b, c, $\ldots$ is written as the number $a + b \cdot 2 + c \cdot 2^2 + \ldots$. Then for the matrices we list their columns as bit-strings so that the $2{\times}2$ matrix $[(1,0), (0,1)]$ is written $(1,2)$ and the other two as $(1,3)$, and $(3,2)$. This proves a convenient shorthand for matrix algebra over Z_2 because $(a,b){\cdot}1 = a$, $(a,b){\cdot}2 = b$, and $(a,b){\cdot}3 = a + b$. However a better insight is that the matrix (a,b) states the two elements that it transforms 1, 2 into. The action on 3 is then the discrimination of these.

The level L_3' has 7 elements, viz. $(1,3)$, $(3,2)$, $(1,2)$, and their discriminations give the following table. (The table has been simplified by omitting brackets and commas so that $(1,3)$ is written as 13.)

	13	32	12	21	01	20	33
13	0	21	01	32	12	33	20
32	21	0	20	13	33	12	01
12	01	20	0	33	13	32	21
21	32	13	33	0	20	01	12
01	12	33	13	20	0	21	32
20	33	12	32	01	21	0	13
33	20	01	21	12	32	13	0

This becomes more transparent by taking a different basis: writing 1, 2, 4 respectively for 13, 32, 12. Then evidently we must also write $1+2 = 3$ for $13 + 32 = 21$, 5 for 01, 6 for 20, and 7 for 33. The table becomes the familiar one for the nym sum (or for the group $C_2 \times C_2 \times C_2$).

0	1	2	3	4	5	6	7
1	0	3	2	5	4	7	6
2	3	0	1	6	7	4	5
3	2	1	0	7	6	5	4
4	5	6	7	0	1	2	3
5	4	7	6	1	0	3	2
6	7	4	5	2	3	0	1
7	6	5	4	3	2	1	0

The dcss of 13 are $[1]$, $[2]$, $[4]$, $[1,2,3]$, $[2,4,6]$, $[1,4,5]$, $[All]$, so the multiplicity of L_3' is 7.

This result for the skeleton guides that for the actual hierarchy. Some of the notation can be taken over. L_2 has two dual pairs of generators and so three dual pairs of elements. As well as 1, 2, 3 we need 1*, 2*, 3* because the construction deals with automorphisms, which therefore preserve duality. We have $T u^* = (T u)^*$, so we need specify only the result on 1, 2, 3. Matrices are no longer available but we can still use (a, b) to state the effect of an automorphism on 1, 2. Because the discrimination operation is neither commutative nor associative, this is not quite sufficient to specify L_3. For example observe that in L_3' the automorphisms $(1, 3)$, $(3, 2)$ take $1, 2, 3$ into $a = 1, 3, 2$ and $b = 3, 2, 1$ respectively. Then their discrimination $(2, 1)$ takes $1, 2, 3$ into $2, 1, 3$ and this is just the discrimination $a b$. This means that the third element need not be specified. This is exactly equivalent to the statement that the automorphisms are non-singular matrices. For L_3 on the other hand $(1, 3)$ takes $1, 2, 3$ into $c = 1, 3, 2^*$ and $(3, 2)$ takes it into $d = 3, 2, 1^*$. The discrimination $(3, 2) \cdot (1, 3) = (2, 1)$ takes $1, 2, 3$ into $2, 1, 3^*$ but this is not $d{\cdot}c = 2, 1, 3$. The specification of L_3 requires stating the effect on 1, 2, *and* 3. There are now more signals. For identity, omitting brackets and commas again, $abc{\cdot}abc = zzz$, which we write Z. There are seven other ways in which automorphisms may be not essentially different. For example, $abc{\cdot}abc^* = zzy$ and so on. Define $yzz = p$, $zyz = q$, $zzy = r$. Then $qr = zyy$, $rp = yzy$, $pq = yyz$, and $pqr = yyy$, which we write Y. The automorphisms p, q, r and their products form the centre of the algebra, together with their squares, the identity Z. This centre is the group $C_2 \times C_2 \times C_2$. The generators of L_3 are 132^*, 321^*, and 123. To see what is happening it is useful to construct the table for L_3 :

	132^*	321^*	123	2^*1^*3
132^*	Z	$2^*1^*3^*$	$z^*1^*1^*$	3^*2^*1
321^*	213	Z	$2z2$	132
123	$z11$	2^*z2^*	Z	3^*3y
$2^*1^*3^*$	321^*	1^*3^*2	33^*y	Z

Taking a different basis as in the corresponding table in the skeleton, $1, 2, 4$ for 132^*, 321^*, 123 and using the signals, this can be written:

	1	2	3	4
1	Z	3	$Y2$	5
2	$Y3$	Z	$r1$	$rp6$
3	2	Y	Z	7
4	$qr5$	6	$pq7$	Z

Evidently instead of elements coming in dual pairs they come in octets. Any two members of one octet are not essentially different. Each octet consists of the product of an element of the skeleton with one of the eight elements of the centre. We have therefore proved that the characteristic group of L_3 is L_3' as promised above. The multiplicity of L_3 is 7. It is now easy to complete the table for L_3, but we shall not need it. The reason it is not needed is the essential uniqueness of L_3.

We complete this section by giving the proof promised above that any dcs will have an automorphism (usually many) to characterize it. It is sufficient to prove this for the skeleton because, if there were a dcs in the actual hierarchy for which there were no automorphisms, this would translate into one for the skeleton. Take as an example the dcs $[1, 2, 3]$ in a system with more than 2 generators, *viz.* 4 and 8. The need is to show there is an automophism T such that $T\,1 = 1$, $T\,2 = 2$, and so $T\,3 = 3$ but for all other elements $T\,u \neq u$. T is specified if its values for all the other generators are given. For the generators 4 and 8 define $T\,4 = 8$, $T\,8 = 4{\cdot}8 = 12$. The values for this automorphism are then:

4	5	6	7	8	9	10	11	12	13	14	15
8	9	10	11	12	13	14	15	4	5	6	7

So evidently $T\,u \neq u$ for all relevant u. If there were more extra generators, say 4, 8, 16 the corresponding trick would be to set $T\,4 = 8$, $T\,8 = 16$, and $T\,16 = 4{\cdot}16 = 20$. If there were only one extra generator, 4, take $T\,4 = 1{\cdot}4 = 5$. It is easy to generalize the proof to any level and any dcss. We have now completed the investigation of the forms of the discrimination operation and we have shown the complete agreement of some of the results with the lowest levels of the Parker-Rhodes hierarchy.

The next stage is to see how the process account differs from the original one given by Parker-Rhodes at the higher levels, and to calculate the effects of these differences on the physical constants that come up.

In the original Parker-Rhodes construction, the bit-strings of length 2 generate 3 dcss. The automorphisms (non-singular $2{\times}2$ matrices) characterising them are unique. These three matrices, as bit-strings of length 4, generate 7 dcss, and the $4{\times}4$ matrices characterising them are not unique. There are 74088 sets of such matrices. Of these, 61160 are sets that generate, as bit-strings of length 16, 127 dcss. We call such sets *normal*. The remaining 12928 sets generate fewer dcss. Parker-Rhodes avoided this difficulty by requiring the mathematician to choose one of the normal sets. Then the 127 dcss would require for their characterisation $2^{127} - 1 \approx 10^{38{\cdot}2}$ $252{\times}256$ matrices. Such sets cannot be normal; Parker-Rhodes' choice becomes impossible and the construction halts. The cumulative totals of dcss when the hierarchy changes level are then 3, 10, 137, and $10^{38{\cdot}2308}$ (from successive sums of 3, 7, 127, and $10^{38{\cdot}2308}$ where $3 = 2^2 - 1$, $7 = 2^3 - 1$, and

$10^{38 \cdot 2308} = 2^{127} - 1$). These four scale constants are direct consequences of the Parker-Rhodes construction.

The evident relation of the last two to the reciprocals of the coupling constants of the electromagnetic field and the gravitational field suggest that the first two are similarly related to nuclear forces (and that there is no 'fifth field' in nature). Here however we confine ourselves to $\alpha = 1/137 \ldots$ since only in this case is the physical constant known very accurately (to 12 significant figures).

If the construction were constrained in some way to remain at the first three levels then $1/137$ would be a probability, *viz.* the probability that a random pair of dcss were the same. The close numerical relation of this probability to $e^2/\hbar$ justifies the identification of this scale constant. We call $1/137$ the *bare value* of α. Such a constraint is not possible in the process construction.

Instead we calculate the probability that the construction remains at the first three levels. For this purpose we assume a generalised ergodic hypothesis — that when there are several possibilities in the process, each will occur with equal probability. At the first level, a dcs may be one of the three at that level or none of them (because it is at a higher level) so each has probability $1/4$. Similarly at the next levels there are probabilities $1/8$ and $1/128$, so long as Parker-Rhodes'choice of a normal set of matrices is made. The probability of being at none of the levels is therefore e where $1/e = 4 \times 8 \times 128 = 4096$, and so the probability that the construction condition happens to be fulfilled is $1 - e$. The probability $1/137$ thus becomes $(1 - e)/137$ giving a revised value for $1/\alpha$ of 137.033447265625 (*i.e.* $137(1 + 1/4096)$).

This value of e has been found by making Parker-Rhodes' choice of matrices and such a choice is ruled out for the process. We now calculate the correct value of e. We begin by doing this for the skeleton hierarchy since this closely resembles that of Parker-Rhodes. Doing this will later guide the calculation for the actual hierarchy. The process uses any of the 74088 sets with equal probability and so there will not always be 127 dcss. The number 128 in the formula for e will be decreased, as will $1 - e$ and so $137.033 \ldots$ will be further increased. A lengthy calculation gives the value of $1/\alpha$ as $137.035104872 \ldots$ (in error by less than 0.001%). The calculation could obviously be carried out as a computer search. This has not been done because the details below are needed in the accurate calculation for the actual hierarchy and that is much less easily computerised.

The calculation falls into two parts: firstly the calculation of what numbers of dcss there can be less than 127; secondly the determination of the number of occurrences of each. These numbers then determine the probabilities by a second use of the generalised ergodic hypothesis. The first part is straightforward. Suppose a set of seven generators $[a, b, c, d, e, f, g]$ fails to generate 127 dcss. Then one element, say g, must be the result of discriminating r of the others ($r = 1, 2, \ldots, 6$). The set $[a, b, c, d, e, f]$ by itself will generate $2^6 - 1 = 63$ dcss. The remaining dcss which involve g will not all be new ones. Any of them involving also $r - 1$ of the members of g, or any involving all of them will be repetitions of some of the original 63. Thus there are in all $(r + 1) \times 2^{6-r}$ repetitions, the second factor arising from those members of $[a, b, c, d, e, f]$ that are not in g. The number of different dcss is then $127 - (r + 1) \times 2^{6-r}$ repetitions:

r	=	1	2	3	4	5	6
number	=	63	79	95	107	115	120

The second part of the calculation is the establishing of the number of sets to each of these situations, say P, Q, R, S, T respectively for $r = 2, 3, 4, 5, 6$ (the case $r = 1$ cannot arise). This involves examining each of the 12928 sets to find the nature of its failure to be normal. This working is set out in full in **Note 1** where the values:

$$P = 216, \ Q = 9000, \ R = 1998, \ S = 1600, \ T = 114$$

are calculated.

Each of the values in the table above has to be increased by 1 before taking the reciprocal, so the replacement for 128 in the formula for e will be X where

$$P/80 + Q/96 + R/108 + S/116 + T/121 + U/128 = 74088/X.$$

Here $U = 74088 - (P + Q + R + S + T) = 61160$. The resulting value of X is $121 \cdot 956006801$. Substituting this for 128 in the first correction gives $137.035104872\ldots$, that is $137/[1 - 1/(4 \times 8 \times 121 \cdot 956006801)]$. It is not surprising that the calculation for the skeleton gives such a good approximation to $1/\alpha$, because the skeleton is a close fit to the actual hierarchy and is based on dcss, which are physically important.

We turn now to the calculation for the actual hierarchy. The multiplicities in the actual hierarchy have been shown to be equal to those in the skeleton so that much of the work in **Note 1** holds here too. The only essential difference is in the actual application of the generalised ergodic hypothesis. In the skeleton hierarchy the generators of the second level, which will produce the dcss, are the matrices $(1, 3)$, $(3, 2)$, and $(1, 2)$. In the actual hierarchy these become $(1, 3, 2^*)$, $(3, 2, 1^*)$, and $(1, 2, 3)$. The first of these could equally have been taken to be $(1, 3^*, 2)$ and the second $(3^*, 2, 1)$, since these are not essentially different (in the sense defined above) and do characterise the first two dcss. There is no such ambiguity about $(1, 2, 3)$. The generalised ergodic principle must be applied to the elements of the actual hierarchy. As a result $(1, 3)$ and $(3, 2)$ in the skeleton must be given twice the probability of $(1, 2)$. The other 4 elements generated by these three are not restricted by having to characterise any des and so each can arise as any member of an octet. The corresponding members of the skeleton must then be given eight times the probability of $(1, 2)$. This change in probabilities entails a corresponding change in the probabilities of the various types of non-normal sets. This change is worked out in detail in **Note 2**. It is most convenient to express the results in terms of a notional set of numbers P', Q', R', S', T' which are the former P, Q, R, S, T weighted to take account of the changed probabilities. Their values to three decimal places are

$P' = 304 \cdot 398$, $Q' = 15288 \cdot 198$, $R' = 918 \cdot 777$, $S' = 2001 \cdot 704$, $T' = 48.999$,

with U' defined as $74088 - (P' + Q' + R' + S' + T') = 55525 \cdot 924$.

Inserting P', Q', R', S', T' into the formula

$$P'/80 + Q'/96 + R'/108 + S'/116 + T'/121 + U'/128 = 74088/X'$$

gives $X' = 118.917238212\ldots$, and so $1/\alpha = 137/[1 - 1/(4 \times 8 \times 118 \cdot 917238212)] = 137.036011393\ldots$. This agrees to better than one part in 10^7 with the value $137 \cdot 035999710\ldots$ reported from Harvard by Gabrielse (Phys.Rev.Lett. 97, 030801, 2006).

The very close agreement between the calculated constant and the observed value of $1/\alpha$ emphasises the correctness of our identification of α as a basic scale constant which can be calculated from the process idea. It is therefore important to consider the small discrepancy. This discrepancy does not arise from an arithmetic slip. The arithmetic has been repeatedly worked over and at the end of **Note 2** we show how robust the calculation is against changes in the counting of sets. It is true that suggestions have been made in the experimental physics community that different experimental conditions may not determine exactly the same constant, but these suggestions seem no longer to be taken seriously. The remaining possibility, and one we have not investigated is this: — the calculation in **Note 2** is of a compound probability arising from two uses of the generalised ergodic principle. There may be some subtlety in this compound probability that we have not taken into account.

NOTE 1.

We are indebted to Ron F. Weeden, who checked and re-calculated these notes and made valuable suggestions to improve the presentation.

This note calculates the values of P, Q, R, S, T in the text. The first step is to construct the sets of operators that give rise to zeros. The second level of the hierarchy has 3 generating elements. Denote these by the bit strings 100, 010, 001 which, in conformity with the numbering system in the text, can more shortly be written as 1, 2, 4. The whole second level, $L = L_2$, then consists of 1, 2, 4, $1 + 2 = 3$, $1 + 4 = 5$, $2 + 4 = 6$, and $1 + 2 + 4 = 7$. There will also be 7 dcss:

$$[1],\ [2],\ [4],\ [2,4,6],\ [1,4,5],\ [1,2,3],\ \text{and}\ D[1,2,4]$$

where DT denotes the discriminate closure of the set T. The most general matrix operators leaving these fixed are respectively:

$$(1,a,b),\ (c,2,d),\ (e,f,4),\ (g,2,4),\ (1,h,4),\ (1,2,i)$$

and the identity $I = (1,2,4)$. We shall denote these by A, B, C, D, E, F and I. Here a, b, c, d, e, f, g, h, i are bit strings, elements of L but not all such are allowed. For consider first the matrix $A = (1,a,b)$. Its effect on L can easily be tabulated:

x	1	2	3	4	5	6	7
Ax	1	a	$1+a$	b	$1+b$	$a+b$	$1+a+b$

Since A is to be a non-singular matrix, $a \neq 0, 1$; $b \neq 0, 1$; and $a + b \neq 0, 1$. But also there are to be no other eigenvectors than 1, so $a \neq 2$, $b \neq 4$, and $a + b \neq 6$. This leaves 14 possible pairs (a,b):

$$(a,b) = 3,6;\ 3,7;\ 4,3;\ 4,6;\ 4,7;\ 5,2;\ 5,6;\ 5,7;\ 6,2;\ 6,3;\ 6,5;\ 7,2;\ 7,3;\ 7,5.$$

Similar considerations apply to (c,d) and to (e,f). Each pair will have 14 possible values so in all these 3 operators can arise in $14^3 = 2744$ ways.

Secondly consider the matrix $D = (g,2,4)$, the effect of which on L is:

x	1	2	3	4	5	6	7
Dx	g	2	$g+2$	4	$g+4$	6	$g+6$

giving rise to the conditions $g \neq 0, 2, 4, 6, 1$. Thus g has only three possible values $3, 5, 7$. Similarly for h, i, so that in all there are $2744 \times 27 = 74088$ sets of matrices, as stated in the text. We have to determine how many of these are normal, that is give rise to 127 dcss and how many are not.

This calculation is simplified if we take advantage of a basic symmetry group. This arises from simultaneous equal permutations of the rows and columns of the matrices. Since the operators are 3×3 matrices this group, of order 6, is the permutation group on three letters. Consider first the interchanging of second and third columns and rows, which we shall denote by P. Clearly a, b, are interchanged by P and are also changed in value. Thus P produces the changes $2 \leftrightarrow 4$ and $3 \leftrightarrow 5$ in the bit strings and so the matrix $(1,3,6)$ for example is transformed into $(1,6,5)$. It is convenient to tabulate the values of a, b in a square table. Here the ordering of the a-list is arbitrary but the b-list has been derived from it by P:

$a\backslash b$	5	2	3	6	7
3	0	0	0	1	1
4	0	0	1	1	1
5	0	1	0	1	1
6	1	1	1	0	0
7	1	1	1	0	0

Here 1 denotes that the pair is allowed. The symmetry of the table has been produced by the deriving of the b-list from the a-list by P.

Next introduce R, the interchange of first and second rows and columns. By R, $(1,a,b)$ is transformed into $(a',2,b')$ where the primes denote the transformed values. To every allowed $(1,a,b)$ corresponds a $(c,2,d)$ where $c = a'$ and $d = b'$. Now R produces the changes $1 \leftrightarrow 2$ and $5 \leftrightarrow 6$ in the bit-strings and so it is clear that the corresponding table for c, d, is the same except that the column for c will read $3,4,6,5,7$ and the row for d will read $6,1,3,5,7$. Then applying P to this new table means transforming $(c,2,d)$ into $(c',d',4)$ so $e = c'$ and $f = d'$ with corresponding ranges of values $5,2,6,3,7$ and $6,1,5,3,7$. It is convenient to combine the three tables into one, and this will be much used in the calculation:

TABLE 1

$e\backslash f$			6	1	5	3	7
	$c\backslash d$		6	1	3	5	7
		$a\backslash b$	5	2	3	6	7
5	3	3	0	0	0	1	1
2	4	4	0	0	1	1	1
6	6	5	0	1	0	1	1
3	5	6	1	1	1	0	0
7	7	7	1	1	1	0	0

A similar use of the symmetry group is that, by applying R to the allowed values $3, 5, 7$ of g we derive $3, 6, 7$ for h and by applying P to these we get $5, 6, 7$ for i. Three other tables, while not essential, prove helpful in the analysis. In investigating the way in which the seven matrices generate the next level we shall be concerned with $a + f$, $b + d$, and $c + e$:

TABLE 2

$a\backslash f$	6	1	5	3	7		$b\backslash d$	6	1	3	5	7		$c\backslash e$	5	2	6	3	7
3	5	2	6	0	4		5	3	4	6	0	2		3	6	1	5	0	4
4	2	5	1	7	3		2	4	3	1	7	5		4	1	6	2	7	3
5	3	4	0	6	2		3	5	2	0	6	4		6	3	4	3	5	1
6	0	7	3	5	1		6	0	7	5	3	1		5	0	7	0	6	2
7	1	6	2	4	0		7	1	6	4	2	0		7	2	5	1	4	0

Here the entries are readily calculated but it is useful to have them displayed.

Turning now to the main calculation of zeros, the seven generating matrices will be a normal set and so give rise to 127 elements at the third level (and therefore to 127 dcss) unless a sum of some of them is zero. We have to determine how such zeros arise. We remark first that the four matrices D, E, F, I form a normal set. This can be seen at once by considering any discrimination between them and recalling the condition $g \neq 1$. The total number of elements at the next level is then at least $4^* = 15$ where $r^* = 2^r - 1$. The other possibilities are 31 or 63 as well as 127 when the whole set is normal. The calculation determines how many sets of each. We find below an efficient way of carrying out this calculation.

It is necessary to have some systematic labelling of the various cases. Zeros can arise from a set whose members are any elements of $\mathcal{D}[D, E, F, I]$

or none of them together with some members of $\mathcal{D}[A, B, C]$. We shall label the four basic members D, E, F, I by the same bit string method adopted above so denoting them by $1, 2, 4, 8$. Then a list of members of $\mathcal{D}[D, E, F, I]$ is

0					8	1	2	4
1	g	2	4		9	$1+g$	0	0
2	1	h	4		10	0	$2+h$	0
3	$1+g$	$2+h$	0		11	g	h	4
4	1	2	i		12	0	0	$4+i$
5	$1+g$	0	$4+i$		13	g	2	i
6	0	$2+h$	$4+i$		14	1	h	i
7	g	h	i		15	$1+g$	$2+h$	$4+i$

The zeros that arise can be divided into three kinds, according as they come from some or none of $\mathcal{D}[D, E, F, I]$ together with 1, 2, or 3 respectively of $\mathcal{D}[A, B, C]$,

FIRST KIND.

Here just one member of A, B, C is involved so there is no loss of generality in taking it as $A = (1, a, b)$ so long as we multiply the number of zeros found by three to take account of the other two cases. Since the first column of A is 1 a zero can arise only in cases 2, 4, 8, or 14. But the conditions $a \neq 2$, $b \neq 4$ rule out 2, 4, and 8, so the only zero here will be when $(l, h, i) + (1, a, b) = 0$, that is, when $h = a$ and $i = b$. Notice that c, d, e, f, g are unrestricted. From $h = a$ it follows that they have the values 3, 6, 7 only and from $i = b$ they can have only the values 5, 6, 7. But from Table 1 these values are further restricted to $a = 3$, $b = 6$ or 7 and $a = 6$, $b = 5$, and $a = 7$, $b = 5$. There are then just four possibilities :

$h = a$	$b = 1$
3	6
3	7
6	5
7	5

The total number of zeros, taking account of the 14 cases of (c, d) and of (e, f) and of the 3 of g will therefore be $4 \times 14^2 \times 3 = 2352$. Taking account of the symmetry, this results in 7056 zeros of this first kind, derived from 4 elements.

SECOND KIND.

Here again it will be sufficiently general to take B and C as the elements outside $D\,[D, E, F, I]$ and then to multiply the results by 3. Now $B + C = (c + e,\ 2 + f,\ 4 + d)$. Attending to the third column, Table 1 shows that $4 + d$ can have only the values $1, 2, 3, 5, 7$ and this then rules out cases $1, 2, 3, 8, 9, 10, 11$ in the list. In the second column, similarly, $2 + f$ can have only values $1, 3, 4, 5, 7$ and so, in addition, cases $4, 5, 12, 13$ are ruled out, leaving only $6, 7, 14$ and 15. These must be tackled separately.

CASE 6   The condition $(0, 2 + h, 4 + i) + (c + e, 2 + f, 4 + d) = 0$ gives the equations :

$$c = e, \qquad h = f, \qquad i = d$$

leaving (a, b) and g unrestricted. Since $c = e$, their possible values are $3, 5, 6, 7$. The values for $f = h$ are $3, 6, 7$ and for $i = d$ $5, 6, 7$. We tabulate the results:

$i = d$	$\mathcal{A}$	$c = e$	$\mathcal{A}$	$f = h$
5		3		6
5		6		3
5		6		7
6		5		3
6		7		6
6		5		7
7		3		6
7		6		3
7		6		7

Here the letter $\mathcal{A}$ between the columns means that the relation "allowed" according to Table 1 holds between the elements on either side. These nine cases must be multiplied by 14 and by 3 giving 378 and then, because of the symmetry, this gives 1134 zeros arising from 4 elements.

CASE 7   Before going on it is desirable to systematise the procedure. We shall illustrate this by tackling Case 7 first by a graphical method and then using this to show how it can be systematised into what we call the *tableau method*.

The equation for Case 7 is

$$(g, h, i) + (c + e, 2 + f, 4 + d) = 0$$

so that $c + e = g$, $h + f = 2$, and $d + i = 4$. If $i + d = 4$ and $i = 5, 6, 7$ then d can have only the values 1 or 3 (for 2 is forbidden) and so $i = 5$ or

7. Similarly from $h + f = 2$ it follows that $f = 1$ or 5 and $h = 3$ or 7. From these values we then get respectively $c = 5, 6, 7$ or $4, 5, 7$, and $e = 3, 6, 7$ or $2, 3, 7$. The reader will find it instructive to represent this by two graphs, one beginning with 5 and one with 7. These exhibit twelve possibilities and since (a, b) is not determined this gives rise to $12 \times 14 = 168$ zeros, or taking account of the symmetry, $168 \times 3 = 504$ each arising from 5 elements.

The graphical procedure is over-elaborate and it is better systematised in the form of a tableau in this way :

DEFINITIONS.

A *tableau* is a rectangular array of up to 9 columns of *sites*, and an indefinite number of rows. The sites in the left-hand column are called *initial* and those in the right-hand column are called *final*. A site may be *filled* by one of the elements $1, 2, \ldots, 7$ or empty. Every non-final site has an *immediate successor* (the next site on its right) and other *successors* (the sites in the next column to its right). If t is a successor of s, we say that s is a *predecessor* of t. Between each pair of columns is one of the symbols $\mathcal{A}$, $1, 2, \ldots$, which state the relation between an element and those elements allowed to fill its successor sites. Here $\mathcal{A}$ again denotes "allowed" according to Table 1 and $1, 2, \ldots$ denote the equations defining the set of zeros under consideration. Any element can fill initial sites. A site may be *active* or *passive* defined thus:

(a) A filled site is passive;

(b) a site is active if and only if it is not the case that its immediate successor is passive.

RULES FOR FILLING SITES.

Initially all sites are empty and so active.

[1] An active initial site is to be filled by the next unused element.

[2] When any nonfinal site has been filled, the next step is to fill its highest active successor with the next allowed element. If all such allowed elements have already been used, then the original site is called *saturated*.

>(In constructing the tableau it is useful to underline the element in a saturated site.)

[3] When a final site has been filled, the next step is to return to its predecessor and repeat 2. If the predecessor is saturated, then return to its predecessor and repeat 2 and so on. The tableau for Case 7 is then

i	3	d	A	c	1	e	1	g	A	f	2	h
5		1		5		6		3		1		7
.		.		.		2		7		5		7
.		.		6		3		5		5		7
.		.		.		.		.		1		3
.		.		7		2		5		5		7
7		3		4		3		7		5		7
.		.		.		.		.		1		3
.		.		.		7		3		1		3
.		.		.		.		.		5		7
.		.		5		6		3		1		3
.		.		.		2		7		5		7
.		.		7		2		5		5		7

Here the equations have been numbered in the order in which they were stated. The column for g need not have been included. The relation between the e and f columns is simply A.

Three remarks should be made about the tableau method.

Firstly, the rules as stated will give a tableau with rather more entries than above but some of them will not reach the final column. Here impossible entries have been excluded beforehand. It is usually convenient to do this and we then have a *reduced tableau*.

Secondly, some cases are of this general form in which the relations form a linear graph :

$$d \;\text{---}\; c \;\text{---}\; g \;\text{---}\; e \;\text{---}\; f \;\text{---}\; h$$

some will be found to have reentrant graphs. An example occurs below and we shall then show how to modify the tableau method to deal with this.

Thirdly, the reader may well be thinking of a way in which tableaux could all be replaced by matrix algebra. This is quite possible but not useful here because the modified calculation for accurate discrimination in **Note 2** will be found to need the tableaux results. What the matrix method does supply is a useful check on the total number of zeros and we have in fact used it for this purpose.

CASE 14   The equation here is

$$(1,\, h,\, i) + (c + e,\, 2 + f,\, 4 + d) = 0$$

or

$$i = d + 4 \quad (1), \qquad h = 2 + f \quad (2), \qquad c + e = 1 \quad (3).$$

As in Case 7, (1) and (2) require $i = 5$ or 7 and $f = 1$ or 5. The reduced tableau is :

i	1	d	$\mathcal{A}$	c	3	e	$\mathcal{A}$	f	2	h
5		1		6		7		1		3
.		.		.		.		5		7
.		.		5		.		.		.
.		.		7		6		1		3
7		3		4		5		.		.
.		.		5		.		.		.
.		.		7		6		1		3

giving 4 possibilities. Since g and (a, b) are undetermined, this gives rise to $4 \times 3 \times 14 = 168$ zeros and taking account of symmetry the total is $3 \times 168 = 504$ each derived from 5 elements. It will be noticed that there are three rows that do not reach the final column and this shows that the tableau could have been further reduced at the beginning.

CASE 15   Here the equation is

$$(1 + g, \, 2 + h, \, 4 + i) + (c + e, \, 2 + f, \, 4 + d) = 0$$

so that

$$i = d \quad (1), \qquad f = h \quad (2), \qquad c + e = 1 + g = 2, \, 4, \, 6 \quad (3).$$

As in Case 14, (1) and (2) require $i = d = 5, 6, 7$ and $f = h = 3, 6, 7$. The reduced tableau can then be constructed. Here and later, however, we shall set the result out in a more condensed form by first stating the headings for the columns and their relations but then simply stating the columns of elements without the room for the relations. The tableau is then :

$$i \quad 1 \quad d \quad \mathcal{A} \quad c \quad 3 \quad 1 + g \quad 3 \quad e \quad \mathcal{A} \quad f \quad 2 \quad h$$

with the table : (see over)

537533

77

5766

47233

77

3633

77

65233

77

657366

3766

75366

3533

77

737533

77

5766

47233

77

3633

77

65233

77

Therefore 23 rows and (a, b) is not involved, so this amounts to $23 \times 14 = 322$ zeros. Taking account of the symmetries, this gives a total of 966 zeros from 6 elements.

It will be obvious that, if an element occurs twice in a column, the successors of each will be exactly the same and this provides some saving of labour in constructing the tableau.

Third Kind.

We come now to the remaining kind of zero, when all three of A, B, C are involved in the form of $A + B + C = (1 + c + e, \ 2 + a + f, \ 4 + b + d)$. It will transpire that all 16 cases can arise (but not all need to be dealt with separately).

CASE 0   Here D, E, F, I are not involved, so the equations are

$$c + e = 1 \quad (1), \qquad a + f = 2 \quad (2), \qquad b + d = 4 \quad (3).$$

A reduced tableau results by noting that these equations imply $c \neq 5$, $e \neq 3$, $a \neq 6$, $f \neq 3$, $b \neq 6$, $d \neq 5$. But notice also that this is one of the cases mentioned above where the graph of the equations is reentrant. The last d may not be possible, given the initial c. So some rows must be deleted. The tableau is :

$$c \quad 1 \quad e \quad \mathcal{A} \quad f \quad 2 \quad \mathcal{A} \quad b \quad 3 \quad d$$

with the table :

325751	x
26	x
37	.
7526	x
62	x
73	x
457526	x
62	x
73	.
676437	.
62	x
73	x
1362	x
73	x
5751	.
26	x
37	.
761362	x
73	.
7526	.
62	x
73	.

Here the x beside certain rows indicates that the row should be deleted because the final d is not consistent with the initial c. The number of rows remaining is then 8 and since g, h, i are not involved, the number of zeros is $8 \times 27 = 216$ derived from 3 elements. It will be observed that a great deal of effort is wasted here in constructing rows which are then forbidden. This will be worse in later cases when the number of possibilities is much larger. We will now explain how to avoid this by what is known as a 'cut-and-paste' tableau, or CPT for short. This will be explained for

CASE 1   Here the equation is

$$(g,\, 2,\, 4) + (1 + c + e,\, 2 + a + f,\, 4 + b + d) = 0$$

so that

$$1 + c + e = g \quad (1), \qquad a + f = 2 \quad (2), \qquad b = d \quad (3).$$

Before calculating this, observe that the permutations P, R, transform this case into Cases 4 and 2 respectively so we need only construct one tableau and multiply the result by 3. For Case 1 a reduced tableau comes by noting that $a = f = 3,\, 5,\, 6,\, 7$ and $b = d = 3,\, 5,\, 6,\, 7$. Moreover $c + e = 2,\, 4,\, 6$ and so from the tables $e,\, c = 5, 3;\, 7, 3;\, 2, 4;\, 6, 4;\, 2, 6;\, 3, 5;\, 7, 5; 5, 7; 3, 7$. This reduced tableau will be found to have a large number of deleted rows and this can be avoided by the CPT method mentioned above. In this method the tableau is cut in half by a vertical cut between the middle two columns. Then the first 3 columns are filled in the usual way but the initial c gives certain final d's and the last three columns are then constructed from the right instead of from the left. Taking the tableau as

$$a \quad 2 \quad f \quad \mathcal{A} \quad e \quad 1 \quad c \quad \mathcal{A} \quad 3 \quad b$$

the cut-and-paste tableau is constructed below. The result is to give 22 rows but since h, i, are free this corresponds to $22 \times 9 = 198$ zeros or, for the three cases of 1, 2, and 4, a total of $3 \times 198 = 594$ from 4 elements.

(for the tableau, see next page)

afe	cdb
335	566
2	7
6	377
.	4
.	6
552	566
3	7
7	377
.	4
.	6
663	355
7	4
.	6
.	433
.	5
.	7
775	355
2	4
6	6
.	433
.	5
.	7

A short explanation of the top section will make the construction clear.
Initially a is given the value 3 so $f = 3$ and possible values for e are then,
from Table 1: 5, 2, 6. Now $a = 3$ means, from Table 1, that $b = 6$ or 7.
Setting $b = 6$ at the top, this makes $d = 6$ of course, and $d = 6$ corresponds,
from Table 1 again, to $c = 5, 7$, whilst $b = d = 7$ corresponds to $c = 3, 4, 6$.
Now we use the list above to see how many (e, c) can be made up. For
example the first section yields $5, 7$; $5, 3$; $2, 4$; $2, 6$; $6, 4$ that is, five in all.
The overall result is 22 as before.

(Case 3 is on the next page)

CASE 3   The equation for case 3 is :

$$(1 + g, 2 + h, 0) + (1 + c + e, 2 + a + f, 4 + b + d) = 0$$

which leads to

$$g = c + e \quad (1), \qquad h = a + f \quad (2), \qquad b + d = 4 \quad (3).$$

Possible pairs (to reduce the tableau) are then :

$c, e = 3, 6; \; 4, 3; \; 4, 7; \; 6, 5; \; 6, 3; \; 5, 2; \; 5, 6; \; 7, 2$

$a, f = 3, 5; \; 4, 3; \; 4, 7; \; 5, 6; \; 5, 3; \; 6, 1; \; 6, 5; \; 7, 1$

$b, d = 5, 1; \; 2, 6; \; 3, 7; \; 7, 3.$

The CPT is then as below, and gives 44 rows.

baf	ecd	
561	561	
5	3	
71	25	
	6	(7)
	27	
256	256	
3	6	
61	27	
5	.	(7)
71	.	
343	637	
7	34	
61	7	(15)
5	56	
71	3	
735	343	
43	7	
7	25	(15)
56	6	
3	27	

Here i is unrestricted, so these correspond to $3 \times 44 = 132$ zeros. But the symmetry shows the same numbers to come for Cases 5 and 6 so that the total is of 396 zeros each from 5 elements.

CASE 7   The equation here is :

$$(g,\, h,\, i) + (1+c+e,\, 2+a+f,\, 4+b+d) = 0$$

so that

$$c + e = 1 + g = 2, 4, 6 \quad (1),$$
$$a + f = 2 + h = 1, 4, 5 \quad (2),$$
$$b + d = 4 + i = 1, 2, 3 \quad (3).$$

Possible pairs are :

$c, e = 3, 5;\ 3, 7;\ 4, 2;\ 4, 6;\ 6, 2;\ 5, 3;\ 5, 7;\ 7, 5;\ 7, 3$

$f, a = 6, 3;\ 6, 7;\ 1, 4;\ 1, 5;\ 5, 4;\ 3, 6;\ 3, 7;\ 7, 3;\ 7, 6$

$b, d = 5, 6;\ 5, 7;\ 2, 1;\ 2, 3;\ 3, 1;\ 6, 5;\ 6, 7;\ 7, 6;\ 7, 5.$

A CPT for

$$c \quad 1 \quad e \quad \mathcal{A} \quad f \quad 2 \quad a \quad \mathcal{A} \quad b \quad 3 \quad d$$

is :

abd	cef		abd	cef		abd	cef	
365	536	1	431	461	3	521	461	4
7	7	1	65	53	2	3	53	3
76	37	3	7	7	2	65	7	3
5	5	1	76	37	3	7	37	3
.	357	3	5	5	2	76	5	3
.	7	1	.	425	3	5	.	.
.	42	3	.	6	4			16
.	6	3	.	53	2			
.	46	3	.	7	2			
		19	.	37	3			
			.	5	2			
					28			

(CPT continued on next page)

(CPT continued from previous page)

abd	cef	
656	353	1
7	7	4
21	42	2
3	6	3
31	46	2
.	357	1
.	7	4
.	42	2
.	6	3
.	46	2
		24

abd	cef	
756	536	4
7	7	4
21	37	1
3	5	4
31	353	1
.	7	4
.	42	2
.	6	3
.	46	2
		25

and $19 + 28 + 16 + 24 + 25 = 112$ zeros from 6 elements. Note that in this tableau there is inserted a right-hand column of numbers. These are the numbers of zeros corresponding to the value of c in that row (the final one being the column's sum).

CASE 8   The equation is :

$$(1, 2, 4) + (1 + c + e, 2 + a + f, 4 + b + d) = 0$$

so that $c = e$, $a = f$, $b = d$. A straightforward tableau is

$f = a$	$b = d$	$c = e$
6	5	3
.	3	7
5	6	7
.	7	3
3	6	5
.	7	6
7	5	6
.	3	5

so there are in all only 8 rows. But g, h, i are unrestricted so this corresponds to $8 \times 27 = 216$ zeros, from 4 elements.

CASE 9 (which also serves for CASES 10 and 12). Here the equation is

$$(1+g, 0, 0) + (1+c+e, 2+a+f, 4+b+d) = 0$$

so that

$$c+e = g = 3,5,7 \quad (1), \qquad a+f = 2 \quad (2), \qquad b+d = 4 \quad (3).$$

Possible pairs are :
  $a, f = 3,1; \ 4,6; \ 5,7; \ 7,5;$
  $b, d = 5,1; \ 2,6; \ 3,7; \ 7,3;$
  $c, e = 3,6; \ 4,3; \ 4,7; \ 6,5; \ 6,3; \ 5,2; \ 5,6; \ 7,2.$
  A CPT is :

abd	cef	
373	361	
.	5	1
.	43	1
.	6	
.	47	1
		3

abd	cef	
437	436	2
73	6	1
.	47	2
		5

abd	cef	
526	657	
73	52	2
.	7	2
.	36	.
.	5	2
		6

abd	cef	
751	525	2
26	7	2
37	43	1
.	6	2
.	47	1
		8

giving 22 rows; here h, i are not involved so the total number of zeros is 198. Together with Cases 10 and 12 this is $3 \times 198 = 594$ zeros from 5 elements.

CASE 11 (which also serves for CASES 13 and 14). The equation is :

$$(g, h, 4) + (1+c+e, 2+a+f, 4+b+d) = 0$$

so that

$$c+e = 1+g = 2,4,6 \quad (1), \qquad a+f = 2+h = 1,4,5 \quad (2),$$
$$b = d = 5,3,6,7 \quad (3).$$

Possible pairs are :
  $c, e = 3,5; \ 3,7; \ 4,2; \ 4,6; \ 6,2; \ 5,3; \ 5,7; \ 7,5; \ 7,3;$
  $a, f = 3,6; \ 3,7; \ 4,1; \ 4,5; \ 5,1; \ 6,3; \ 6,7; \ 7,6; \ 7,3,$

(for the tableau, see next page)

A CPT is :

baf	ecd	
563	535	3
7	7	1
76	24	3
3	6	3
.	26	3
		13

baf	ecd	
341	243	4
5	6	4
63	35	3
7	7	3
76	57	3
3	3	3
		20

baf	ecd	
636	356	4
7	7	4
41	57	1
5	3	4
51	.	.
		13

baf	ecd	
736	537	1
7	7	4
41	24	2
5	6	3
51	26	2
		12

showing a total of 58 rows. Here i is not involved so this corresponds to $3 \times 58 = 174$ zeros and with those of Cases 13, 14 we have $3 \times 174 = 522$ zeros from 6 elements.

CASE 15   The equation is :
$$(1+g, 2+h, 4+i) + (1+c+e, 2+a+f, 4+b+d) = 0$$
so that
$$c+e = g = 3,5,7 \quad (1), \quad a+f = h = 3,6,7 \quad (2), \quad b+d = i = 5,6,7 \quad (3).$$
Possible pairs are :

$a, f = 3,5; \ 4,3; \ 4,7; \ 5,6; \ 5,3; \ 6,1; \ 6,5; \ 7,1;$
$b, d = 5,3; \ 2,5; \ 2,7; \ 3,6; \ 3,5; \ 6,1; \ 6,3; \ 7,1.$

A CPT is :

abd	cef	
361	525	3
3	7	3
71	43	1
.	6	2
.	47	1
		10

abd	cef	
436	653	3
5	52	4
61	7	4
3	36	1
71	5	4
.	657	3
.	52	4
.	7	4
.	36	1
.	56	4
		32

abd	cef	
525	436	3
7	6	4
61	47	3
3	653	4
71	52	3
.	7	3
.	56	3
.	36	2
		25

(CPT continued on next page)

(CPT continued from previous page)

abd	cef			abd	cef	
653	361	3		753	361	3
25	5	2		25	5	2
7	43	4		7	43	4
36	6	3		36	6	3
5	47	4		5	47	4
.	525	2				16
.	7	2				
.	43	4				
.	6	3				
.	47	4				
		31				

This gives rise to 114 zeros from 7 elements.

. . .

(continued on next page)

We now list these results:

				Totals
3 elements				
	THIRD KIND	(0)	216	216
4 elements				
	FIRST KIND		7056	
	SECOND KIND	(6)	1134	
	THIRD KIND	(1, 2, 4)	594	
		(8)	216	9000
5 elements				
	SECOND KIND	(7)	504	
		(14)	504	
	THIRD KIND	(3, 5, 6)	396	
		(9, 10, 12)	594	1998
6 elements				
	SECOND KIND	(15)	966	
	THIRD KIND	(7)	112	
		(11, 13, 14)	522	1600
7 elements				
	THIRD KIND	(15)	114	114
				12928

The right-most column gives the values of P, Q, R, S, T in the text.

NOTE 2.

This note is a continuation of NOTE 1 and uses the same notation. The calculations for the skeleton will be modified to give the figures for.the actual hierarchy. As explained in the text, the elements of the second level have weights (unnormalised probabilities) proportional to :

Element	1	2	3	4	5	6	7
Weight	1/4	1/4	1	1/8	1	1	1

These weights affect the probabilities of occurrence of the operators A, B, C, D, E, F. To understand this, recall the set of zeros labelled FIRST KIND in NOTE 1. Here there are four possible cases of A (*i.e.* of (a, b)) : 3,6; 3,7; 6,5; 7,5. There are 14 possibilities for A and in the skeleton each was taken to be equally likely and so to have probability 1/14. There is therefore a probability 4/14 that one of the four possible cases arises. But the rest of the calculation proceeds with numbers of zeros rather than probabilities, so the probability is multiplied by the number of possibilities, 14, to give the number 4. In the actual hierarchy the weights of the 14 possible pairs (a, b) are : 1 for the eight cases not involving 2 or 4 (1 does not occur), 1/4 for the three involving 2, and 1/8 for the three involving 4. The operators D, E, F contain only elements of weight 1 so that their rôle in the actual hierarchy is exactly the same as in the skeleton and so they can be disregarded in this Note. The total weight for (a, b) is 9·125 and the probabilities are no longer 1/14 but 1/9·125, 1/36·5 and 1/73 respectively. In the FIRST KIND zeros the elements 2, 4 do not arise so each of the four possible cases has probability 1/9·125. This is again multiplied by the number of possibilities, 14, giving 6·13698630137. This is the suitably weighted number of notional zeros to be used in the rest of the calculations. We write N for these notional numbers, so here $N = 6·13698630137$. Under the symmetry transformation $1 \leftrightarrow 2$ A is transformed into B and so the factor $14/9·125 = 1·53424657534 = J$ say, applies equally for (c, d) if B is in play. It is not quite the same for (e, f). Here the table of possible pairs can be written down from the table of allowed values in Note 1 :

5,3; 5,7; 2,5; 2,3; 2,7; 6,1; 6,3; 6,7; 3,6; 3,1; 3,5; 7,6; 7,1; 7,5.

As well as the eight with weight 1 the remaining six are all of weight 1/4 giving a total of 9·5. The corresponding factor $14/9·5 = 1·47368421053 = K$ say will also be needed. In the later calculations we shall also need the numerical values $J^2 = 2·35391255395$ and $J^2 K = 3·46892376372$.

In the FIRST KIND case no elements of lower weight arise so the final N is greater than the actual number in the skeleton. As an example of the opposite consider SECOND KIND Case 7 in Note 1. The number 12 was found there from a graph. What is needed is to trace each final element back to see whether 1, 2, or 4 (which we shall call *critical elements*) occurs and to note the result. The top 7 in the graph has a 1 so its weight is 1/4. The next 7 has both a 2 and a 1, so its weight is 1/16. The twelve weights can be seen to total 1·5625. Both (c, d) and (e, f) are involved so this weight must be multiplied by $J K$ giving 3·53280461428 in place of 12. Thus 168 in the calculation becomes 49·4592645999 (*i.e.* 168 × 3·53280461428/12). For no ambiguity the graph needs lines put in, so it would be better to work from the tableau. These are the general principles for the revision of the calculation to fit the actual hierarchy. There are certain other specific technical features which will be explained as we proceed through all the cases in the same order as in Note 1.

FIRST KIND.

In the calculation for the skeleton the total 2352 was multiplied by 3 to take account of the other cases arising from symmetry. Here the calculation for a, b will give $6\cdot13698630137 \times 3 \times 14^2 = 3608\cdot54794521$ and the same result for c, d. For e, f the corresponding figure is $(14/9\cdot5) \times 4 \times 3 \times 14^2 = 3466\cdot10526316$ giving a total N of **10683·2011536** rather than the 7056 for the skeleton, from 4 elements.

SECOND KIND.

CASE 6

Here again critical elements do not occur in the table and so the 378 cases have to be multiplied by JK in the case tabulated and again by symmetry when c, d is unrestricted but by J^2 when e, f is unrestricted. Now $J(2K + J) = 6\cdot87590246022$ so that $N = $ **2599·09112996** from 4 elements.

CASE 7

The case when (a, b) is not determined was calculated above giving $N = 49\cdot4592645999$. When (c, d) is not determined, symmetry gives the same N. The remaining case when (e, f) is not determined is found by using the original graph but changing the nodes by the symmetry transformation $1 \leftrightarrow 4$. Equivalently we can use the original graph but note that each 4 produces weight 1/4, as with each 2, but 1 produces 1/8. The resultant weights total 1·40625. Multiplying by J^2 gives 3·31018952899 and $N = 46\cdot34265340589$. The three cases together give $N = $ **145·2611826056** arising from 5 elements.

CASE 14

In the tableau of Note 1 the 4 weights total 0·625. Multiplying by JK gives 1·41312184571 and so $N = 59\cdot35111751983$. One symmetry transformation gives the same but the other $(1 \leftrightarrow 4)$ gives a total weight of 0·28125 and so $N = 27·80559204354$ by the same method as in Case 7. The three cases total $N = \mathbf{146·50782708319}$ from 5 elements.

CASE 15

Here the tableau for (a, b) not determined has 23 rows, 12 of which involve 2 or 4. These 12 have a total weight 1·625 giving a total weight of 12·625. Multiplying by JK and by 14 gives $N = 399·63085796683$ in each of two cases. In the third case the symmetry transformation gives a total weight of 13·25. Multiplying this by $14 \times J^2$ gives $N = 436·65077875774$ and so $N = \mathbf{1235·91249469141}$ from 6 elements.

THIRD KIND.

CASE 0

The 8 rows of the tableau have total weight 3·25 and multiplying by $J^2 K \times 27$ gives $N = \mathbf{304·39806026607}$.

CASE 1

Here the 22 cases were exhibited in a CPT. This is a particularly simple CPT for our purposes because the critical elements 2 and 4 all occur along the cut. Consider the first section :

e	c
335	566
2	7
6	377

Here the allowed joins across the cut are when $c + e = 1 + g = 2, 4, 6$. They are therefore 5,7; 5,3; 2,4; 2,6; 6,4, with weights 1, 1, 1/32, 1/4, 1/8 making a total of 2·40625. The other three sections can be treated similarly giving 4·28125, 4, and 2·5625 respectively, with an overall total of 13·25. Multiplying this by $9 K J^2$ gives $N = 413·66915882312$. Symmetry produces another case of the same value and a third one with 1 and 4 interchanged. The overall total in this case is 13·75 giving $N = 429·27931575984$. The three cases together give $N = \mathbf{1256·61763340609}$ from 4 elements.

CASE 3

This is the more general CPT in which some critical elements occur away from the cut. We set out the details for the first sub-tableau to make the method clear. The method is to list the elements on each side of the cut, each with the weight arising from the three elements leading up to it. Then one must find the corresponding weights when the two sides are joined. To save printing a forest of fractions and to aid the calculation we list the partial weights in the form of powers of 2, i.e. 2 for $1/4$, 3 for $1/8$ and so on :

$$
\begin{array}{cccc}
w & f & e & w \\
2 & 1 & 5 & 2 \\
0 & 5 & 3 & 2 \\
2 & 1 & 2 & 4 \\
. & . & 6 & 2 \\
. & . & 2 & 4 \\
\end{array}
$$

Now possible f, e pairs are (from the Table) $1,6$; $1,3$; $5,2$; $5,3$ of which the first three occur twice each. The weights (in index notation) therefore add to $4, 4, 4, 4, 4, 4, 2$. The total weight is thus $6/16 + 1/4 = 0.625$. The remaining sub-tableaux give another with weight 0.625 again and two others of weights 2.875. This can be calculated by a technique which we shall call *expanded* CPT (or ECPT). We explain this first by re-doing the first tableau above. This *generates* an ECPT in the following way. Taking the first line, if $f = 1$ then e must be 6, 3, or 7 from Table 1 so that the ECPT begins :

$$
\begin{array}{cccc}
2 & 1 & 3 & 2 \\
. & . & 6 & 2 \\
\end{array}
$$

To save labour it is useful to note that there is another 2 1 on the third line, so this would all be repeated. We denote this by $\star 2$ on the right :

$$
\begin{array}{ccccc}
2 & 1 & 3 & 2 & \star 2 \\
. & . & 6 & 2 & \\
\hline
\end{array}
$$

Then $f = 5$ requires $e = 2$, 3 or 7 so the tableau continues

$$
\begin{array}{cccc}
0 & 5 & 3 & 2 \\
. & . & 2 & 4 \\
. & . & 2 & 4 \\
\hline
\end{array}
$$

The complete ECPT for this first tableau is therefore:

$$
\begin{array}{ccccccc}
2 & 1 & 3 & 2 & \star 2 & 4 \\
. & . & 6 & 2 & & 4 \\
\hline
0 & 5 & 3 & 2 & & 2 \\
. & . & 2 & 4 & & 4 \\
. & . & 2 & 4 & & 4 \\
\hline
\end{array}
$$

where the right-hand column gives the weight for that line in index notation. This gives rise to a 2 and six 4's so to the total weight of 0·625.

Before passing on to the other three sub-tableaux it is useful to consider the way in which symmetries can be used to give Cases 5 and 6. The ECPTs for the other cases, which can be generated by the symmetries $1 \leftrightarrow 4$, $2 \leftrightarrow 4$ respectively will have the same form but when 1 and 4 are interchanged the effect is the same as that of leaving the CPT unchanged and instead giving the weight of any 1 as 3 (in index notation) and of any 4 as 2. Then the first sub-tableau would become :

$$
\begin{array}{ccccccc}
3 & 1 & 3 & 3 & \star 2 & 6 & \bullet \\
. & . & 6 & 3 & & 6 \\
\hline
0 & 5 & 3 & 3 & & 3 \\
. & . & 2 & 5 & & 5 \\
. & . & 2 & 5 & & 5 \\
\hline
\end{array}
$$

giving $1/8 + 2/32 + 4/64 = 0·25$.

Returning to Case 3 it is now clear how to proceed for the remaining three sub-tableaux and the complete ECPT is :

(for the tableau, see next page)

w	f	e	w	Tot	
2	1	3	2	4	*2
.	.	6	2	4	
0	5	3	2	2	
.	.	2	4	4	
.	.	2	4	4	
2	3	2	2	4	
.	.	6	0	2	
.	.	2	2	4	
4	1	6	0	4	*2
2	5	2	2	4	
.	.	2	2	4	
3	3	6	0	3	*2
.	.	5	0	3	
2	1	6	0	2	*2
.	.	3	3	5	
.	.	7	3	5	
.	.	3	0	2	

w	f	e	w	Tot	
0	5	3	3	3	
.	.	7	3	3	
.	.	3	0	0	
0	5	3	3	3	
.	.	7	3	3	
.	.	2	2	2	
.	.	2	2	2	
3	3	2	2	5	*2
.	.	6	0	3	
.	.	2	2	5	
0	6	3	3	3	
.	.	7	3	3	
0	3	2	2	2	
.	.	6	0	0	
.	.	2	2	2	

The total weights are then, in index notation, two 0s, ten 2s, twelve 3s, twelve 4s and eight 5s (a total of 44 in agreement with the original calculation) so a total weight of $2 + 5/2 + 3/2 + 3/4 + 1/4 = 7$. To calculate the Case when 1 and 4 are interchanged, *i.e.* 1 is given weight 3 and 4 weight 2 we introduce a more compact way of setting out the ECPT. We exhibit this with the first sub-tableau.

w	f	e	w	
3	1	5	2	–
0	5	3	2	636
3	1	2	5	5
.	.	6	2	66
.	.	2	5	5

Here it is most convenient to organise the work by the right-hand set of rows. For the top row $e = 5$ requires $f = 3$ or 7, which are not present on the left-hand side of the cut, so there is a 'dash' against 5. The next row, $e = 3$ corresponds to $f = 1$, 5, or 6 so to weights 3, 0, 3. These are to be added to the right-hand weight 3 giving 6, 3, 6 and so on. The whole tableau is then :

3	1	5	3	–
0	5	3	3	636
2	1	2	5	5
.	.	6	3	66
.	.	2	5	5

2	6	2	2	44
2	3	6	0	255
5	1	2	2	44
2	5	.	.	
2	1	.	.	

2	3	6	0	3322
2	7	3	2	525
3	1	7	2	525
0	5	5	0	22
3	1	3	0	303

0	5	3	2	22
2	3	7	2	22
2	7	2	2	2244
0	6	6	0	220
0	3	2	2	2244

with total weight $2 + 17/4 + 5/8 + 8/16 + 8/32 + 4/64 = 7\cdot6875$. The interchange $1 \leftrightarrow 2$ will necessarily give the same result so the three Cases 3, 5 and 6 give a total weight of $22\cdot375$. Multiplying by $3J^2K$ gives $N = \mathbf{232\cdot85150763943}$ from 5 elements.

(Case 7 is on the next page)

CASE 7

The technique just described yields the following tableau, which has been shortened by noticing that $c = 5$ and $c = 7$ give the same results :

0	5	5	0	0000
0	7	3	0	000000
0	6	4	5	555
0	5	6	2	222
.	.	4	3	333

5	1	4	5	888888
3	5	5	2	757575
3	7	3	2	555
3	6	3	0	333
3	5	.	.	
.	.	6	2	7555
.	.	5	0	535353

4	1	4	5	7555
2	3	5	2	642642642
0	5	3	2	222
0	7	.	.	
0	6	.	.	
0	5	.	.	

0	6	3	0	00
0	7	7	0	04220422
4	1	4	5	5757
2	3	4	3	3535
2	1	6	2	264264

0	6	5	0	4220422042204220
0	7	3	0	00
4	1	4	5	57
2	3	4	3	35
2	1	6	2	264

giving total weight $20+24/4+12/8+12/16+24/32+6/64+8/128+6/256 = 29 \cdot 1796875$ so that $N = \mathbf{101 \cdot 22211138655}$ from 6 elements.

CASE 8

There are no critical elements here so multiplying by $8 J^2 K$ gives $27 \cdot 75139010973$ and so $N = \mathbf{749 \cdot 28753296264}$ from 4 elements.

(Case 9 is on the next page)

CASE 9

An ECPT is :

0	3	5	2(3)	2(3)
.	.	4	5	5
.	.	4	5	5

0	3	5	2	2
.	.	7	2	2
.	.	5	0	0

(2)3	7	4	3(2)	6(4)
.	.	6	0	3(2)
.	.	4	3(2)	6(4)

(3)2	1	5	2	4(5)
.	.	7	2	4(5)
.	.	6	0	2(3)

(2)3	3	4	3(2)	6(4)
.	.	4	3(2)	6(4)

2	6	5	2	4
.	.	7	2	4

2	6	5	2	4
.	.	7	2	4
.	.	5	0	2

0	7	4	3(2)	3(2)
.	.	6	0	0
.	.	4	3(2)	3(2)

Ignoring the bracketed numbers gives $2+5/4+3/8+6/16+2/32+4/64 = 4{\cdot}125$. Multiplying by $9\,J^2\,K$, $N = 128{\cdot}78379472795$. The same number arises for Case 10 since this is generated by the $1 \leftrightarrow 2$ symmetry. Case 12 results from the $1 \leftrightarrow 4$ symmetry so again the same ECPT can be used with a change of weights. These changed weights are the numbers in brackets and give $2+6/4+2/8+8/16+4/32 = 4{\cdot}375$. Hence $N = 136{\cdot}58887319631$. In all the three cases give $N = \mathbf{394{\cdot}1564265222}$ from 5 elements.

(Case 11 is on the next page)

CASE 11

An ECPT is :

0	3	5	0	000
0	7	7	0	0
0	6	2	4	444
0	3	6	2	222
.	.	2	2	222

0	6	3	0	523052305230
0	7	.	.	
5	1	5	0	0
2	5	.	.	
3	1	.	.	

5	1	2	4	4444
2	5	6	2	7222
0	3	3	0	520520520
0	7	.	.	
0	6	5	0	000
0	3	.	.	

0	6	5	0	0
0	7	7	0	5320
5	1	2	4	46
2	5	6	2	257
3	1	2	2	24

with a total of $16 + 18/4 + 4/8 + 8/16 + 8/32 + 2/64 + 2/128 = 21 \cdot 796875$. There are two such cases and one of $20 \cdot 265625$ so a total for the three of $63 \cdot 859375$. Multiplying by $3\,J^2\,K$ gives $N = \mathbf{664 \cdot 56991042064}$ from 6 elements.

(Case 15 is on the next page)

CASE 15

Finally the ECPT for this Case is :

d	c			
2	1	5	2	424424
0	3	4	3	33
2	1	6	0	22

3	6	6	0	355355
3	5	5	2	577557755775
5	1	5	0	35533553
3	3	3	0	33
5	1	.	.	

2	5	4	3	355355
2	7	6	0	22222222
2	1	5	2	424424
0	3	5	0	202
2	1	3	0	22

0	3	3	2	244
2	5	5	2	222222
2	7	4	5	57755775
0	6	4	3	35533553
0	5	6	2	442
.	.	6	0	220

0	3	3	2	442
2	5	5	2	22
2	7	4	5	57755775
0	6	6	2	442
0	5	.	.	

The total weight is 14·125 which multiplied by $J^2 K$ gives
$N = $ **48·99854816249**.

(The listed results are on the next page)

We now list these results as we did in Note 1 :

				Totals
3 elements				
	THIRD KIND	(0)	304·398	304·398
4 elements				
	FIRST KIND		10683·201	
	SECOND KIND	(3, 5, 6)	2599·091	
	THIRD KIND	(1, 2, 4)	1256·618	
		(8)	749·288	15288·198
5 elements				
	SECOND KIND	(7)	145·261	
		(14)	146·508	
	THIRD KIND	(3, 5, 6)	232·852	
		(9, 10, 12)	394·156	918·777
6 elements				
	SECOND KIND	(15)	1235·912	
	THIRD KIND	(7)	101·222	
		(11, 13, 14)	664·570	2001·704
7 elements				
	THIRD KIND	(15)	48·999	48·999

with an overall total of 18562·076. Here the values have all been rounded to three decimal places. The right-most column gives the values of P', Q', R', S', T' in the text, with U' defined as $74088 - 18562·076 = 55525·924$.

Although empirical adequacy cannot be taken as an indicator that a theory is correct, the most important prerequisite for the correctness of a theory is empirical adequacy. That the calculated value of $1/\alpha$ agrees with the experimental one to better than one part in 10^7 fulfils this strikingly.

That the calculated value of $1/\alpha$ agrees with experiment to better than one part in 10^7 is a strong confirmation that the theory is a correct one. The remaining discrepancy (at present 0·000011683) requires further discussion. In this note we only show now that it is not caused by rounding errors. For consider a change p in P' (because P' comes in the formula with the smallest denominator). Then U' changes by $-p$ and so $74088/X'$ changes by $p(1/80 - 1/128)$. In all then $1/X'$ changes by $6\cdot3 \times 10^{-8} p$. From the formula the change in $1/\alpha$ is roughly $(137/32)(6\cdot3 \times 10^{-8})p = 2\cdot7 \times 10^{-7} p$. A rounding error in P' represented by $p = 0\cdot001 = 10^{-3}$ then produces a change in $1/\alpha$ of only $2\cdot7 \times 10^{-10}$.

Ron F. Weeden informs us that this value of $1/\alpha$, which must be a rational number from the derivation, is

$$363354665990215680/2651526867264751.$$

The numerator is $137 \times 73^2 \times 29 \times 19 \times 11^2 \times 5 \times 3^6 \times 2^{11}$. The denominator has four prime factors, $7 \times 1901 \times 230309 \times 865177$, whose significance is opaque.

Chapter 7

Process and iteration

Contents

Contrast of conventional prescription by laws and the sequential picture. Process has no sudden jump at which there is physical interpretation. Interpretation is there from the beginning and builds up with the mathematical equipment available. In process one has to start independent constructions or duplicate them. Time first and space later and less fundamental. The levels incompatible with all-or-none interpretation. Pace Parker-Rhodes. He didn't realise his own levels demanded a new philosophy. Significant numbers characterise the levels. In fact cumulative summation entailed process. The numbers come from the discrimination algebra and are only derivatively connected with space structure. The changes in level enable contact to be made with conventional physical theory provided that they can be identified with the idea of a physical field. The increase in complexity of successive levels has to go with increase in descriptive power of the fields. In fact the most important increase is the introduction of electromagnetic fields. The assignation of the coupling constants to fields is done elsewhere.

In the course of this book a picture emerges in which a scheme of algebraically generated numbers dictates the whole shape of theoretical physics. Here we make this principle explicit and definite and look at its most important consequences. These numbers are connected to physical theory by an initial identification with particles through the fine-structure constant approximately $1/137$, and this number is central to the algebraic scheme that we start from. Further development of the scheme will bring in other coupling constants. Of course it is important that this number and its refinements occur with the correct status in the world. In the rest of the book we shall explain how this status is justified in two straightforward cases: high-energy particle theory in Chapter 8 and the structure of space-time in Chapter 9. The more tangled message about non-relativistic quantum mechanics is taken up in Chapter 11. The requirement about the status of the algebraic construction is, however subsidiary to the recognition that there should be such numbers imposed, as it were, upon physical theory. This is the vital point: they have to be *imposed*: they do not emerge from the practice of physical method in the way that properties of particles are usually assumed to arise in conventional physics. The "imposition", as we call it, arises from the algebra of discrimination that gives centrality to the quadratic group, and its non-commutative refinement as was described in the foregoing chapter.

Our picture of the world comes by repeated applications of the central algebra, and for that reason we may speak of a *process theory*. The repetition is conveniently conveyed by the term *iteration*.

Before we discuss the contrast between conventional physics and the process approach we must explain an important distinction that arises in the process. For the sake of simplicity, attention was not drawn to this in Chapter 6 because the calculation carried out there arises most clearly in the special situation described. In general, however, a decision has always to be made between two possible procedures. In using the convenient shorthand description as decision, it is not intended to imply that the system is interfered with or that someone makes a decision. Decision simply means that one or other procedure is followed in the process. When two elements come into play and produce a third, the three constitute a dcs (a discriminately closed set). Suppose a putative new element comes into play. It is then possible for the system to go up a level. In Chapters 5 and 6, for the sake of simplicity, we assumed that it did go up a level. In this case a hierarchy would be generated and would lead to the calculation of the fine-structure constant. Now, going up a level is one of two possible procedures. We will call it the *upwards construction*. The alternative procedure, which we shall call the *transverse construction*, arises because it is not possible to proceed by discriminating the new element with each of the three existing ones in turn. This can lead to an infinite regress, as can be seen as follows: Consider the simplified case of the skeleton and suppose that the new element is in fact one of the original ones. One discrimination has obviously a probability $p(1) = 1/3$ of giving a signal. In the remaining 2/3 of cases an additional element is produced, so that a second discrimination has a probability $p(2) = (2/3) \times (1/4)$ of yielding a signal. The general $p(n)$ is $2/(n+1)(n+2)$ and so the average number of discriminations to produce a signal is the sum of $2n/(n+1)(n+2)$. This series is easily seen to be divergent. We are left with only one possibility: to leave the existing dcs as it is and start again. In due course another new element comes in and so the new ones will generate a new dcs. The new structure thus arising consists of pairs of elements; one from the original set and one from the new one, (a, b). This is the transverse construction.

We turn to the contrast between our construction and conventional physics. To most physicists it is quite clear how bodies of theory predict the numerical results of experiments. The theory — including its mathematical structure — resembles aspects of the world. For example the continuum of real numbers may correspond one-to-one with the points on a line that we traverse or the succession of times registered by a clock in our experience. Then it may happen that a number that arises in the theory agrees to some extent with the measured value of the corresponding thing in the world. This will be called a prediction, and the theory in question has a degree of credibility that is given to it through the success of these predictions. What goes on in the world is described as far as possible by a set of 'laws' and these take a mathematical form. The possible movements of things in a successful theory are exactly what is laid down in these mathematical formulations. The mathematics is therefore said to *govern* the motions of the bodies or particles. A very simple example will show what is meant by 'governing'. Any free body such as a planet moves in an ellipse or other of the conic sections and so the mathematical expression that described the motion may be $x^2/a^2 + y^2/b^2 = 1$. If we take some numerical value for x say $x(1)$, then the expression will dictate a corresponding value for $y(1)$ giving a point, $x(1), y(1)$ on the plane.

If we ask a physicist why bodies are constrained to follow the curves specified by equations most will say that that is because the equations have been chosen to formulate Newton's laws of motion. However, this answer is at best a short cut that leaves out a lot of relevant history. Galileo's foundational enquiries were based on the concept of a body that must move in a straight line except when in the presence of some separate influence or force. There is therefore only one thing we know about it. To say that its motion deviates from the straight line is only to say that we know something additional about the motion. To say that the motion is governed by an equation says that we know an infinite number of things about the motion. The foregoing chapter gives a very different picture in which there are numbers that stand outside this prediction process and are essential to the structure of the whole body of theory.

To make the picture of the last chapter comprehensible in the terms of the conventional point of view it is necessary to explain how there could be physical content in the algebraic construction of the last chapter. We have been insisting that certain numbers generated from our mathematical principles are the essential, and indeed the only, way to identify the scheme with physical theory. These numbers appear as the successive bounds to a sequence of 'levels' together with the numbers that characterize them; and it is very natural to see them as the origin of coupling constants (as is argued in detail in different ways in other chapters). The most conspicuous of these coupling constants is the fine-structure constant α. The statistically approximate value of α is $1/137$, or $1/(3 + 7 + 127)$. We can divide the task of justifying this identification into two. We have

(a) to clarify just what is physically essential about the construction,

(b) to justify interpreting 137 as an approximation to $1/\alpha$.

Here the main emphasis will be on (a), and the answer this book proposes is that a process view is needed. There are two immediate reasons for introducing the idea of process. The first is the need to express algebraically the construction of successive new sets of entities (in level change). The second reason is to throw light on their progressive nature. There is not just one Combinatorial Hierarchy. The original construction started with two elements and these give rise to three dcss which uniquely determine three elements at the next level. At this next level these three elements give rise to seven dcss, and these seven do NOT uniquely determine seven elements at the next level. In fact there are 74088 possible sets of elements at the next level. Moreover at this next level these seven elements can give rise to 127 or to 63 or 31 or 15 dcss and in each case there are very many sets of next level elements.

In the constructive process account we never get a complete spatial picture: we always have to start again each time in the process. The conventional idea that mathematical expressions govern and therefore represent and therefore explain the motions of bodies does away with the need to say any more about how or why things happen in sequences. In that case there is no need to look any further into the causation of the process that is going on. However our method is to enquire what it is that the mathematical progression was meant to encapsulate in abstraction from the mathematical details, and this enquiry is what our concern with process is all about.

If the process view is to be adequate for the weight we are going to be putting on it we shall have to elaborate exactly what we mean by process. It will not be enough to point out that a world without process would be without past or future, nor to point out that present-day physics describes the world as comprised of interacting entities rather than of static Newtonian masses. The notion of process has been around for a long time, and we begin by looking at three examples of the idea among philosophers to see how they refined and elaborated the notion. Consider Plato. Notably in the Timaeus, there is a cosmology built on a general notion of changing inter-connectedness and that is Plato's idea of process. Plato's concern here is to refute Heraclitus' insistence on everything being in a state of flux. Heraclitus, according to Plato, said "Nothing ever is, everything is becoming". The whole Platonic Universe is process in this sense, but with the additional notion of retention of connectedness which Plato sees as constituting the Receptacle. The Receptacle is Plato's way to establish something permanent in the Heraclitean flux. We mentioned above the Platonic Receptacle as being at variance with our notions. We should mention here that what is wrong with the Receptacle from our point of view is not that it is an assumption of space. Plato himself warns that it is a more complicated thing. Our objection is rather that Plato's notion is of a complete finished universe arising through the process. Plato has a second defence against the universality of the flux. This second defence comes in a different context, notably in the Republic. It is the theory of Forms. We shall not need anything corresponding to Forms, but they are not inconsistent with our view. Indeed Heisenberg has been quoted as seeing elementary particles as Platonic forms.

Plato's idea of process is an ontological one. With Kant on the other hand it is principally a matter of epistemology. The change results from his criticism of Plato's reliance on reason alone but the underlying epistemology mirrors Plato's ontology. For example, Kant is initially concerned to distinguish between sensations and concepts and his reason for wishing to do so is to preserve the unstructured continual flux of sensations. The concepts in terms of which the sensations can be understood reflect Plato's search for permanence. Cognition is "awakened into exercise by means of objects that affect our senses" and this view of cognition is an epistemic analogy of the ontology of the Heraclitean flux. Kant's response is to find permanence in the synthetic *a priori*. Initially he is content simply to point out the existence of such propositions as "Space has only three dimensions". The necessity of having such assurance of permanence comes later when Kant considers such matters as the freedom of the will. As with Plato, Kant has a second line of defence which is his system of categories.

Whitehead is the modern philosopher usually connected with the notion of process. Whitehead sees process as a succession of changing relations between "things": but "thing" is not to be read here in the everyday sense. Perhaps it is best interpreted to mean whatever can be related by relations. Such an interpretation is close to our notion of elements coming into play. For Whitehead as for Plato and Kant there is a need to qualify the process in order to account for order in the world. So he assumes a realm of enduring forms which he calls "potentialities". These are not part of the process. The necessity for permanence is more severe for Whitehead than for Plato or Kant because they were content to see changing incompleteness of our knowledge of the world as a barrier to understanding that would vanish when completeness reigned: for Whitehead it is this changing character that is needed for knowledge to be possible. His emphasis on incompleteness agrees with our objection to Plato's Receptacle.

Evidently Whitehead's notion of process is closer to ours than is Plato's or Kant's. Whereas Plato shows forms attaining permanence, Whitehead sees that "Each actual thing is only to be understood in terms of its becoming and perishing." This is more in line with our notion of elements coming into play and being deleted. To sum up, what these philosophers have in common is that each sees the need for some element of permanence in the process — the Receptacle for Plato, the *a priori* for Kant, and Whitehead's potentialities. They also see a need for considerable detail about the actual operation of the process and out of this detail comes their second line of defence against lack of permanence. Our notion of permanence is an abstraction from all three. Elements of the system are constructed in a sequence; we speak of elements coming into play. Discrimination sometimes has the effect of deleting elements, consistently with Whitehead's views. The need for permanence is equally important for us and is provided for in two ways. The first is by iteration.

In the conventional picture there are also changes that appear against a stable background, but the changes are to be identified with our *observations*. The changes are what we observe; and so in that picture observation is the way what we call iteration proceeds. For us there is no level at which observation appears. It is there from the beginnings of our algebra, and to use a once very common term there is nothing that is *uninterpreted*.

Let us see how these different points of view appear in some actual physical cases. In macroscopic physics, for example, observation requires a particle that has the attributes of mass, charge and therefore both gravitational and electromagnetic reactions to what are then called fields. This particle is an idealization that enables us to imagine what is going on in terms of the behaviour of a fixed and invariable constant and stable thing. We may call it a *test particle*. It is an idealization because it is assumed not to disturb the situation it is being used to investigate. It satisfies our requirement of being the thing that can be run time and time again.

When we go to quantum physics, the fixed and invariable stable thing whose behaviour lets us see the structure of the world has a different guise but it is still there. We cannot now speak of test-particles although any particle could be called a test particle; they are what the world is made up of. It is then no longer possible to assume the non-disturbance of the system. Physicists would still speak of 'observation' because that is just commonsense talk but in fact 'observation' has become just a special sort of discrete process in a theory that is essentially discrete. It is a restricted class of the things that go on in the world — namely those things that happen to have been observed. The error in using 'observation' as the trick on which the state/observer philosophy depends is fully discussed in Chapter 11.

In the picture we are presenting there are different categories of observable things, and these have different degrees of permanence. Our picture has to be contrasted with the current view according to which there are entities that have a high degree of permanence which distinguishes them from other things that may be observed. These are the fundamental particles of physics, (m, e etc.) and those with less permanence are made up out of them.

The second provision of permanence which arises naturally out of the system comes from the scale constants. In this respect our notion of process is more akin to Plato's where his forms arise from the system, and to Kant's where the same is true of his *'a priori'*. It is less akin to Whitehead's thinking since he had to introduce his 'potentialities'. Of course there has always been active speculation about the dimensionless ratios that may be formed out of the particle attributes and what sense to make of the even higher degree of permanence that their being formed as ratios which are pure numbers seems to entail. For us there is no puzzle and that higher permanence is exactly what we should expect since we have claimed to deduce the most permanent entities — namely the dimensionless numbers. We have the converse problem. We have to explain how entities with lesser degrees of permanence arise without resorting to giving them a distribution in space.

It is useful to recall how it became evident that one couldn't simply introduce spatial language and concepts as a matter of commonsense as Parker-Rhodes had done. He based his mathematics upon the need to express algebraically the construction of successive new sets of entities out of the operations upon the elements of a previously existing set. Because of this progressive construction, the process idea was already implicit since one had to acquire these sets in some order. With a view to physical application two courses were open. One could regard the construction process as a once for all thing, or one could see it as happening over and over again. The latter is the view we take as the one necessary to implement the process idea.

From the outset it was clear that we needed to relate the numbers of physically effective things in the stages of the construction with important physical numbers. Parker-Rhodes drew our attention to the numbers thrown up by his construction, having already noticed their possible physical application to the coupling constants. However the suggestiveness of the numbers only appeared if one formed cumulative sums of them, and he did nothing to justify that procedure. He thought of the appearance of numbers with a physical application by analogy with the eigenvalues of quantum theory. It became progressively clearer that justification of the formation of cumulative sums required the process principle, and Parker-Rhodes probably never took that idea seriously. He found the appearance of 3 and 4 a sufficient justification for linking the algebra with space and time and followed current thinking in being untroubled about conflating two such incompatible sets of ideas. However it was not at all in his mind to fiddle about mathematically to procure a particular sequence of special numbers. That suggestion would be a travesty of his thinking.

The 4 elements at level 1 had originally been constructed through study of the dimensional structure of space-time. However we were trying to abstract from the spatial ideas, and as a final result came to understand the constructed combinations of elements as *discriminately closed subsets*, (see Chapter 5 for the detailed argument). By inspection at an early date we found that to get the numbers right for experimental identification it was necessary to add those of the different stages together. This empirical necessity meant that the second of the meanings for the process idea just given was unambiguously the right one. If the successive levels of construction were just elaborations of the one before, then it made no sense to add them up.

The development of the mathematics from sequential principles and its identification of characteristic numbers with coupling constants always went hand in hand with an attempt to interpret that development in normal physical terms. On the second of these two fronts it always seemed important to link the change from one level to the next more complex one with an increase in the complexity of what can be described. As one goes up a level one changes ones theoretical power. In Chapter 2 we spoke of 'theory-languages', and a change in the descriptive power with change of level is a change of theory-language. If we assume that the simplest theory-language is that exhibited by the non-metrical structure of classical mechanics then one expects that the next level will correspond to electromagnetism since the electromagnetic behaviour of a test-particle is more complicated than that of a particle acting under purely mechanical forces. Thus an increase in descriptive power comes about.

The clue to the connection of sequential principles with physical language is to be found in the principle that in the hierarchical construction the structure at any one level is the basis for that at the next higher level. That principle can once more be imagined physically by specifying fields in terms of the behaviour of a test particle. If the test-particle experiences an acceleration in the direction of its motion, we attribute that acceleration to the presence of a field that is purely mechanical. (It may be gravitational.) If the field is in another direction (that is to say with a component at right-angles to the first) then we need new conceptual apparatus to describe it. We say we are in the presence of an electric or a magnetic field. In continuum language we pass from $\text{grad}V$ to $\text{curl}A$, but we express that change in terms of the process theory. The changes due to the placing of new levels in the hierarchy algebra are thus given a new interpretation using the whole familiar language of classical electromagnetic theory.

It is important to realise that our calculation of the fine-structure constant (with the other couplings) has already introduced the essentials for development of electromagnetism in principle. What we are doing here is to explain how the usual ideas of fields come about. We emphasize that we are having to obtain the concepts without the usual metrical space, and that way of working requires an important departure from what is usual.

We have given central importance to the fine-structure constant in the interpretation of our algebra. Now we must link that interpretation to its place in physical theory. The historical route to it from the quantum theory of around 1920 gave rise to the simple view that of course there is an adequate experimental meaning for that quantity upon which more refined theories may be built because it is just a parameter like any other. The difficulty with that view is to see how one can interpret that constant in itself (and without the help of considering it as a ratio with an electromagnetic constant) as arising as an interaction strength.

One controversial explanation of the fine-structure constant invokes the anthropic principle and argues that the value of the constant is what it is because stable matter and therefore life and intelligent beings could not exist even if the value were different by even a small amount. For instance, were α to change by 4%, carbon would no longer be produced in stellar fusion. If α were greater than 0.1, fusion would no longer occur in stars. This argument is totally at variance with our approach.

The fine-structure constant was originally introduced into physics in 1916 by Arnold Sommerfeld as a measure of the relativistic deviations in atomic spectral lines from the predictions of the Bohr model. Historically, the first physical meaning for the fine-structure constant, α, was the ratio of the velocity of the electron in the first circular orbit of the relativistic Bohr atom to the speed of light in vacuum.

It is important to get a picture of it that will explain why it has to be there. It helps to see it as the quotient between the maximum angular momentum allowed by relativity for a closed orbit and the minimum angular momentum allowed for it by quantum mechanics. It appears naturally in Sommerfeld's analysis and determines the size of the splitting or fine-structure of the hydrogen spectral lines.

There is another way of looking at the fine-structure constant which connects physics with our calculations more closely. Pierre Noyes[8] modified an argument of Dyson[9] which had shown why the renormalised perturbation series in QED cannot be convergent. Noyes followed Dyson in interpreting $1/\alpha$ as a count, in this way: consider a number of charges e and mass m in a clump spaced apart by their Compton wavelengths. The Coulomb energy of each is e^2/r where $r = \hbar/m$ (taking $c = 1$). For N such charges a rough approximation gives a total energy of $Ne^2/r = Ne^2m/\hbar$. This total energy is greater than m and so is enough to produce another such particle if N exceeds $\hbar/e^2 \approx 137$. That is, $1/\alpha$ is a count of how many particles can exist like this. The introduction of iteration leads at once to statistical considerations.

The meaning of 'statistical' and 'random' need to be discussed. Our position is that these terms always refer to the state of our knowledge. If we use a random variable, that is because we do not know the causation, and the expression 'random' warns us against giving misleading information about the extent of what we actually know. This use of 'random' differs from its use in the current observer/state philosophy where 'random' is an absolute choice among mathematical techniques. They (the state/observer theorists) have to say that because otherwise there would be possible alternatives to their way of explaining discreteness. Hidden variable theorizing would no longer be ruled out.

The Born interpretation was originally introduced as purely epistemic: (ψ^2) represents a probability which describes knowledge of the system. This interpretation is now rejected because of Gleason's theorem and the Kochen-Specker result. However these impossibility theorems depend crucially on the use of the real (and complex) fields (see A. Meyer[10]) so they do not restrict our interpretation of statistics, which is like that of the original Born interpretation.

Chapter 8

Pre-space: High-energy particles: Non-locality

Contents

High energy particle theory logically prior to space and time and to non-relativistic quantum mechanics. Classification by symmetries, the quark model entirely consistent with the process algebra (transverse construction). Charge and spin. The weak interaction.

Before considering the first of the two straightforward justifications of the status of our algebraic scheme, we should say more about what form this justification might take. The algebraic structure came from the most abstract possible analysis of physics. It would be naive to expect to see it clearly exhibited in any of the conventional theories. Instead one has to ask the question: is the theory consistent with the framework determined by the structure? We shall find that the two straightforward cases require different techniques to explore this question.

In this chapter we shall introduce as much as is possible about high energy particles without using a background space. The usual way of working would be to establish the principles of quantum theory using the background space and to set the high energy particles within that space. Our general approach requires us to invert this procedure. We have started with a combinatorial structure that we have built up from our basic hierarchical principles and we show that this has the potential to provide a good understanding of the particles. Space appears further down the line. Starting, as we do, with dimensionless numbers, discrete particles will appear before there is a space to put them in. The jump from a combinatoric view of particles to ordinary dynamical concepts can be seen in the Einstein-Rosen-Podolski paradox which is commonly described in terms of a disintegration

95

of a spinless particle producing two leptons of opposite spin. It resulted in the amazing discovery that whatever spin is detected in either determines what must be discovered as the spin of the other. The property of having a particular direction of spin is, according to the conventional quantum mechanics view, not even possessed by the lepton until it has been measured. The new combinatorics recalls that when two shoes are separated the finding of the left shoe instantly determined that the other is a right one.

This discovery is now usually said to show that the relationship between the particles is *non-local*. That is to say it transcends the limitations of spatial connectivity. We invert the argument and say that far from room having to be found for the non-local connectivity, it is that connectivity that we start from. It is the first step, paradoxically, to building a space — which will of course be a relational space. Spatially, the separated particles start as one event, but when their separation is recognised the way is open to create a multiplicity of events, which gives space. It is this possibility that we shall exploit in the construction of space-time, and so of space, in Chapter 9.

Non-locality must come before any details of quantum theory, so it comes in here. Conway (personal communication) has stressed the immediate experimental transparency of the non-local discrete interactions in the tradition of the Einstein-Rosen-Podolski paradox and how they do not depend on any elaborate formalisation. Bowden (personal communication) stresses the necessity of starting from non-locality. "Non-locality is an essential feature of the *standard mathematical formalism* of QM. Anyone who does not understand this understands neither QM nor the experimental results: if it were not for non-locality the world would be essentially classical" (our italics). However, we shall argue in Chapter 11 that this standard mathematical formalism, though an efficient means of calculation for certain phenomena, encapsulates an inconsistency that is fully explained by our theory. The inconsistency is made the reason for the state/observable formalism. That formalism emphasises the amazing nature of non-locality.

Our theory postulates a background consisting of elements that take part in the discrimination process, and in this way generate new elements. There is nothing we can know beforehand about the distribution of these background elements since everything has to come from the construction that is taking place. In particular the background may have the potential for developing spatial ideas but these do not exist yet. We find a strikingly similar picture to be required by particle physicists who postulate a vacuum that is quite independent of physical space. We are thus justified in identifying our background with the vacuum, though unlike the particle theorists we do not have in our minds that there is a physical space there really; to be talked about, as it were, in a different tone of voice. For us, physical space is to be built up progressively by elaboration of the interactions generated from the vacuum. It is really remarkable that the vacuum idea, with its similarity with our background, has been found necessary by the particle theorists.

The main point about the vacuum idea is that it enables physicists to think about particles that are not spatialised. The end of the road towards the particle proves to be the quark. In current physics it is thought puzzling that quarks have no form of spatial appearance as though that was some fault of theirs. By contrast the same limitation arises in the process theory as an essential and inevitable part since it follows from the construction from discrimination. We shall see below that the quarks had to be invented because of the success of the **SU3** basis and of the related higher groups. Quarks are therefore the ultimate development of the concept "particle". People who try to find reasons for "quark confinement" as the failure of the quarks to come with spatial trappings is termed, are putting the cart before the horse.

The quarks arise from juggling the experimental knowledge of combinations of particles. They are precursors of the particles, but it is going in the backwards direction to think of them as things of some material sort banging about inside a nucleus. We do not need to ask how many quarks there are REALLY? as though there were some way of getting behind the combinatoric specification. Current writers seem to find it unhelpful of the quarks to be so shy of exhibiting themselves in experiment, and even try to find explanations for their being that way. If there could be such an explanation then that would show our approach was wrong. The experimental circumstances that quarks would need to manifest themselves in the usual ways do not exist at our present stage of construction.

One can trace the vacuum idea back to the zero-point energy. This name refers to the residual $1/2\hbar$ of action that cannot be eliminated. The history of the famous factor needed for the spin starts from low energy spectral theory and we deal with it elsewhere. There we take apart the different strands that are usually lumped together in the historical account to achieve an unambiguous starting point. Thus interactions in the vacuum gave rise to small but exceedingly accurate spectral displacements at low energies called the Lamb shift and this view would lead to the more familiar picture.

We see the vacuum as the background in our process. However for most physicists the vacuum is full of entities. An isolated charge interacts with the vacuum to produce virtual bosons by emission or absorption. When another charge is present there is a two-way process of bosons emitted by one boson and absorbed by the other. The vacuum acts as a virtual field. (We owe this description to P.Rowlands.) We work, on the other hand, with a hierarchy theory whose initial success is the provision of a scheme of interaction strengths (and therefore coupling constants) with the amazing difference in scale of the electromagnetic and the gravitational constants, and the exactness of the value obtained in the electromagnetic case. To speak of 'interactions' is to presuppose that there are corresponding particles, but this is an ultimately bare concept of 'particle', and later it is necessary to show (in Chapter 9) how the spatial dynamics that is usually thought to come automatically with that concept is built up explicitly.

The discrimination process generates the lowest level of the hierarchy. As we saw in Chapter 6, that level consists of 6 elements which come as three pairs. If we look for such a structure in the high energy domain we find the six quarks and these come as 3 pairs. This correspondence suggests that our theory is on the right lines as will be further argued below.

We should mention here the immensely important appearance of the weak interaction. There really ought to be something that suggests its existence in our algebra, even if we are not able to give much detail. The complexity of the weak interaction is evidenced by the non-conservation of strangeness in some weak interactions, conservation of strangeness or change by 1 in others. We return to this briefly at the end of the chapter; the situation remains unsatisfactory.

The high energy problem has developed in a specific way because of the discovery over the years of hundreds of hadrons. That meant that the emphasis was on classification. There have been several approaches, but we shall concentrate on that based on symmetries. That has been very successful and it gives rise to the quark model which is also the approach nearest to the algebraic hierarchy. We must go into a little detail of the symmetry approach so as to understand the basic notions behind the mathematics. It is most useful to trace the early stages of development before the "standard model" was set in tablets of stone. In any case the standard model depends on a rather large number of disposable constants put in by hand.

The earliest symmetry success arose by noting that the mass difference between proton and neutron was small (0.4%) so that they might profitably be seen as two states of one particle. This "symmetry" was captured by the group **SU2**, that is, the group of unimodular unitary 2×2 matrices. This was the introduction of "isotopic spin" or "isospin". These matrices are essentially the Pauli spin matrices, and it is usually said that isospin, as its name suggests, is a symmetry analogous to electron spin. Such a notion may have played a part historically but it is misleading. It is better to note that the Dirac algebra has the direct product structure **SU2×SU2**. Roughly speaking one of those **SU2** factors corresponds to the non-relativistic electron spin and the other to the electron-positron symmetry. It is the latter which provides the analogy with the proton-neutron symmetry.

Closer inspection of the **SU2** symmetry raises puzzles. One may ask: what exact use is being made of the 2×2 unitary matrices? The answer is that they have just two eigenvectors. If real, as when the matrix is chosen Hermitean, these are 1 and −1, if complex, i and $-i$. So a two-component wave function is envisaged representing a nucleon. Whether the nucleon is a proton or a neutron depends, according to the state/observable philosophy, on a measurement being made. However this is just an unnecessary importation of the state/observable trick. Unnecessary, because all that is required here is the essential two-ness, and this is provided in our process theory by the two elements of Z_2.

Another puzzle is the need for what is called "broken symmetry". The notion of a symmetry group came from geometry. There one considers for example a square in the plane. Rotation through a right-angle gives a square indistinguishable from the original. Such rotations form the cyclic group of order 4. In the physics analysis however, proton and neutron are not indistinguishable though similar in some (important) ways. One says the symmetry is a broken one. Since **SU2** yields rather little, the next step is to consider **SU3**. We start with a little of the experimental evidence. Attributes of particles which are called quantum numbers in low energy physics have a new status when we deal with particles at high energies. We will call them descriptors. Each represents the presence or absence of a property such as having charge or no charge. It is clearest to think of such numerical values like −1, −2, 1/2 as combinations of all-or-none descriptors. The usual aim in high energy physics is to characterise each descriptor by its place in some algebraic structure and then to see this algebraic structure as superseding the classical meaning and definition even though it is assumed there will be a natural progression from the one to the other. (That is an assumption that needs critical examination.)

There is one glaring exception to all this — mass. A very large body of experimental knowledge relates to the masses of the different particles, and particles differ in the amount of it that each possesses, even though particles of the same sort have exactly the same mass. Particle mass is thus like the classical mass in that it can take a range of numerical values. However, it is different in not being able to move continuously between these values. Essentially additivity has been introduced.

By the early 1960s the hadrons had been characterised by 4 internal quantum numbers — charge, baryon number, isospin and strangeness, as well as by two external ones — spin and parity. The conservation of these numbers permitted only the observed scattering events. The next step is to study the masses of particles. A striking example here is the set of 10 particles known by the early sixties that had spin 3/2 and various values of strangeness. There are four non-strange Δ-particles of masses near to $1236 MeV$, three Σ particles whose strangeness is -1 with masses about $1385 MeV$, two Ξ particles of strangeness -2 and mass $1530 MeV$, and one, the Ω^- with strangeness -3 and mass $1672 MeV$. The linear mass-strangeness relation had been shown by 1968[11] to be within 2%. Other groupings of eight, or nine, particles were also apparent. The idea was then to use group structures to exhibit such groupings.

The attraction of this technique for physicists depended partly on the theorem of Lie, that a group was fully described by giving its commutators. The commutators could be studied by considering only the infinitesimal operations in the group. The resulting algebraic structure could then be put in a form already familiar to physicists as creation and annihilation operators. For **SU3** any infinitesimal is $U = I + i e B$, and then the unitary condition (neglecting e^2) is

$$U U^\dagger = (I + ieB)(I - ieB^\dagger) = I + ie(B - B^\dagger) = I.$$

So $B = B^\dagger$, that is B is Hermitean. Moreover, the determinant of U is 1, which easily implies that the trace of B is zero. Hence B has $3^2 - 1 = 8$ degrees of freedom, and the group has correspondingly 8 infinitesimal generators. Two, only, of these can be simultaneously diagonalised; this restriction was identified as corresponding to two conserved quantum numbers. As well as the regular representation of dimension 8, **SU3** has also one of dimension 3. Clearly the "right" numbers were coming up.

Other symmetry groups, **SU6** and **SUn** for various n, later proved useful for classification, but we will stay with **SU3**. The **SUn** system has some predictive power in pointing to patterns of particles with one not observed. The not observed one could then be sought; there were such successes. What is missing is any explanation of why the (broken) symmetries hold. This lack of explanation is tied up with the anomalous and not understood position of mass amongst the particle attributes. Physicists have not ignored the logical problem constituted by the difference between mass and other descriptors. The question of how mass is generated has inspired the hunt for the Higgs boson. We say more about this later.

One of the spin 1/2 octets described by **SU3** consists of the proton p, neutron n and six other hadrons Σ^+, Σ^0, Σ^-, Λ, Ξ, and Ξ^0. In 1996 Gell-Mann and Zweig proposed a new way of looking at these symmetries. This was not a complete explanation because it still left mass unexplained but it is a fruitful change. They proposed that the symmetries arose because the hadrons were "made up of" three of a set of more elementary entities — the quarks. At that time it was thought that the number of quarks would be that of the basic subgroup of **SU3** which has 3 elements. These were given the names up, down, and strange. What was most surprising was that these entities had fractional charges (*i.e.* $-e/3$ for d and s; 2/3 for u). Considerable ingenuity enabled Gell-Mann and Zweig to assign quantum numbers to the quarks so that the proton $p = uud$. Interchanging u and d gives $udd = n$; interchanging d and s on the other hand gives $uus = \Sigma^+$. Not only did the quarks demonstrate why the symmetries held but they opened the way to explanation of the broken symmetries by assigning different masses to different quarks. Such an explanation would also need something about binding energies, so it was far from complete.

Later, the number of quarks was found to be six, in three pairs;- d, u; s, c; b, t (introducing "charm", "bottom" and "top"). The first member of each pair has charge $-e/3$; the second $2e/3$. In order to produce spin 1/2 hadrons the quarks had also to be spin 1/2 entities. Here there comes a naive appeal to non-relativistic quantum theory; — indeed to the old quantum theory. Physical attributes were given to the quarks as if they were the complete particles of ordinary theory. So, it was argued, whatever might be the mysterious way in which the hadrons were "made up of" three quarks, still Pauli's exclusion principle must hold. One of the successes of the quark model had been to exhibit the Ω^- as sss, so as to have strangeness -3, charge $-e$ and spin 1/2, by taking one s with spin $-1/2$ and two with spin 1/2. This led to the introduction of "colour". The two spin 1/2 quarks would have different colours and so would conform to the Pauli principle. There were in all three colours called red, blue and green and in sss all three quarks would have different colours, which balanced to give an uncoloured Ω^-.

A further appeal to quantum theory implies that to each quark corresponds its anti-quark. The anti-quarks have anti-colours. Just as the earlier quark theory has exactly the structure of the lowest level of the hierarchy with six elements grouped in pairs, so the advent of colour and anti-quark also leads to the same algebraic structure: six colours grouped in pairs (red, anti-red, and so on). It is evident that the transverse construction described in Chapter 7 has taken place and that the elements are the thirty-six pairs (a, b). For example $(s, r-)$ would denote the anti-quark of the red, strange quark. This identification of the quark model with the combinatorial hierarchy is very satisfactory. We do not claim that every detail of the model, with the various quantum numbers, comes directly from the hierarchy. What does come is the main framework, or, using the language of Chapters 2 and 4, the theory-language. Thus we can certainly claim that, insofar as the quark model justifies the observed **SUn** symmetries, so our algebraic structure does the same.

One might ask why the results of a transverse construction is the relevant one here? That appears to be a matter of choice by the quark people. If an upward construction had arisen, the next level of the hierarchy would have had 56 elements which would come as 7 octets. That might well provide an alternative theory-language for quarks and hadrons with greater descriptive power.

We turn to what particle theory has to say about how hadrons are formed of three quarks. This again depends on colour which is now thought of as a kind of strong charge. There are 3 colours, and 8 combinations of colours called "gluons" which are thought of as the photons of the strong force. There is only one type of photon in electromagnetism, whereas, since the gluons carry colour between quarks, there are 8 different types of gluons, and they themselves carry colour, so, unlike photons, they can scatter off themselves, even without quarks around. All attraction and repulsion between different like-coloured quarks are thought of as analogous to electromagnetic behaviour between charges. Seeing colour as analogous to charge and gluon to photon, provides an analogy with electromagnetism, though this simple correspondence has the complication that the gluons are attracted to each other as they are being passed between quarks, so that exchange forces are brought in. This is what makes the strong interaction so complicated.

Throughout this statement one continues to see a continual attribution of physical properties to the elements of the process as though these elements are already the complete particles of ordinary theory. We really have to start from a barer picture in which the particles are characterised combinatorially, and we find that this is possible. Consider the eight gluons. There are 8 of them because that is the number of unordered pairs of the six colours and anti-colours which are independent as is seen when the sum of three colours giving zero is taken into account. The number is simple combinatorics and does not depend on experiment. In the combinatorial hierarchy of this book one gets the same thing though as part of a larger structure and without the naive appeal to the old quantum theory. This simple picture would be altogether confused by saying the particles have these numerical values as an experimental fact.

We have thus explained how the combinatorial structure that emerges from high-energy experimentation is that of quarks. An argument that had considerable influence in the early development of the ideas linked the quarks with charge. The quarks were in the first place tripartite. One needed two entities of opposite charge to provide three possibilities, and hence the quarks. You could have two together to give the neutral K^0 meson. You could have two like that and then add a third to give the charged proton if the third addition was the positive one. (The possibility that you could get a negative particle charge-wise symmetrical with the proton and therefore presumably the electron does not seem to have been used, presumably because people thought that the fact that they were dealing with heavy nuclei precluded that. The possibility of bringing in the anti-proton was not available at that date.) The vital thing that emerges from this bit of history is that the appearance of a concept identifiable with charge was needed to get the quark combinatorics off the ground.

The history of the quark model parallels that of the hierarchy in a curious way. In Parker-Rhodes' original construction the lowest level has 3 elements: this is the pair of elements and their discrimination, forming the non-identity elements of the quadratic group. In the correct construction this is modified to 6 elements in 3 pairs, in just the way that the quark model developed.

We have shown, then, that the quark structure is consistent with the algebraic framework. In the remainder of this chapter we shall investigate to what extent the algebraic structure can be related to other aspects of particle theory. We shall initially limit our horizons by sticking with the stable particles with the idea that from there we can find the combinatorial germs of charge and spin. (Not the full concepts with all their trappings — charge replete with blue sparks and so on — though we shall have to show that those trappings follow by a natural progression as we go to higher levels. Just the germs for the moment.) It may sound sensationally ambitious to find these germs of charge and spin in our combinatoric scheme but we have no alternative since we cannot take them as common-sense ideas that emerge from classical continuum physics. We have to have some picture rather than none. Afterwards we shall consider mass, exchange forces and then return to the puzzle of the weak interaction.

As always it is vital to give some idea of how to connect the combinatorial picture with conventional physics. The activity in the skeleton described by the Parker-Rhodes algebra can be presented using the bit-string representation where the progressive discriminations among the bit-strings at one level can be studied by looking at each place one at a time. This activity is permitted by the symmetric difference operation (discrimination) in contrast to what would happen if the operation were, for example, Boolean addition under which, statistically, each place would fill up with and activity would cease. In the discrimination case the activity in one place can be zero if all the bit- strings in play have zero in that place. Strictly the same is the case if all the places are occupied by 1, since 0 and 1 are symmetrical in the symmetric difference operation. However it seems natural to think of the zero as representing absence of activity. To justify this asymmetry in use of the 0, 1 symbols we have to look in the rigorous treatment. This number of places 'in play' is called the Hamming number. For our simple physical picture the Hamming number becomes pretty well the first thing that strikes you. If you are thinking of a lot of bit-strings buzzing around and interacting by discrimination then you are stuck with that number of bit-strings that contain at least one 1. You never can generate ones in a bit-string having occupancy zero. You cannot get a foot in. It was for this reason that John Amson suggested replacing the vector-space model by a projective geometry one.

Our task in this book is to use our combinatoric algebraic picture so as best to bridge the gap between combinatorics and the conventional physics to .which it is meant to be relevant. It appears inevitable in that case to build up the possibility of physical interpretation by asking how many numbers are needed to specify first charge, and then spin. The answer we have given in this writing has been two for charge and one more, three, for spin. At the primitive stage these numbers are indeed all we can assert about these concepts. Interpretation is thus progressive.

A view held by Pierre Noyes was a formative influence in the hierarchy work. He stressed that the counting of interactions was the real entrée into experimental knowledge. Our success with the coupling constants depends on the exact form of the algebraic construction of the hierarchy, and so we must find *in that* the germs of the familiar classical concepts. We cannot start with the assumption that the classical world is really there in the background providing them. This assumption still underlies quantum mechanics since it is assumed there that we already have clear ideas of length, mass, momentum and the rest, even if those ideas have to be modified.

To get a fuller picture we should look again at the status of the calculated numbers. It is reasonable to think of numbers that arise through *counting* of processes (like collision cross sections) rather than fields in specifying classes of particles. At this point we come in with our numbers. Assume for the sake of argument that the calculation of those has been carefully scrutinised and is solid. They have to be pure numbers to be counts and to be independent of dimensional units. However their being pure numbers does not preclude their having *any* sort of physical interpretation. It is not a paradox that they appear in a picture of the world: only that that picture does not have to be dimensional. The picture into which they fit is of an unknown background (though not of course a spatial background) out of which entities come to be part of the construction through a binary operation called discrimination against entities that have already been accrued.

The pure numbers that we seek to interpret as coupling constants are in the first place thought of through the way they afford ratios of the strengths of the fundamental fields. However if there were a definite and universally agreed starting point for the range of constants at the strong interaction level with a small number like unity to characterise it then this number could form the *ratio* 137. As it is, no such thing exists and yet a value, precise to 7 significant figures, is calculated for that number, and no explanation of it as the starting point of the *sequence of ratios* can be given. So far we can say that the strengths of the electromagnetic and gravitational fields are indeed the ratios of the corresponding calculated numbers; but that is not the whole story because the coupling strengths are currently seen as characterising the individual interactions.

To understand how any constant comes to have a value independently of its being a ratio with other such constants we have to start with the electromagnetic constant, about $1/137.036$ or $e^2/\hbar c$. We may try to summarise a complex argument by saying that e is the quantum of charge, and $\hbar$ the quantum of angular momentum, so this ratio provides a comparison of a description in terms of an essentially electrostatic linear field with one in a descriptive language that includes angular momentum. That is why that constant, seen as a ratio, is so prevalent everywhere. It fits very well with our scheme of things to make a clear distinction between descriptive languages with, and without, angular momentum. The fine-structure constant calculation can then be seen as calculating a ratio. With this identification as the starting point, one can advance to the enormous ratio to the gravitational interaction strength and need not demand to interpret that large number as an independent ratio in its own right. The situation with the strong coupling is more complex and more obscure. The practitioners insist that there is a coupling constant but give a variety of values for it. Some give the value 1 and some, values around 10. There is a presumption that the electromagnetism provides a model for it so that it can be some sort of ratio. There is probably no agreement on how that model should be interpreted though, in the sense of what quantities appear as a ratio. Our value is 10, and there seems scope for prediction and clarification at this point.

The other germ is spin. A careful scrutiny of the history has led us to reject the common view that spin cannot be discussed experimentally but has to be regarded as an abstract algebraic construct. However it has to be experimentally immediate. The two-valued character of spin in experiments of the kind first discussed by Einstein, Rosen and Podolsky is absolutely repeatable and comprehensible as an experimental fact that stands clear of any theoretical discussion whatever including the origins of quantum mechanics. When two fermions have a common origin as though they originate in the disintegration of a spinless boson then we find that if one is found to have one value of spin then the other always is found to have the opposite one. This dual appearance is absolute and totally reliable experimentally: the only knowledge the viewer has to have is how to set up the experiments, and therefore this splitting has to be the first thing in our description of the world. It does not require any theory of any sort.

Goudsmid and Uhlenbeck jumped to the conclusion that they were dealing with the addition of a classical angular momentum of the electron; to be faced with Heisenberg's question "Where is the factor two?" Our picture is that this 'germ' of spin appears necessarily in the appearance of the levels in the hierarchy. At the 2-level there are two entities: at the next more complex level there are elements that can handle both. When one such element appears it causes what we see as the two levels being in play as a single action. The one-and-two particle is born, and this is what we have come to call spin.

On the other hand the abstract algebraic concept is not independent of classical angular momentum. This is why spin is of particular importance in showing such a connection. The corresponding case of mass is still puzzling and will need special treatment.

A promising way forward, connecting in the combinatorial approach with the conventional ones, is to be found in work of Kauffman[12] who bases his discrete outlook on the appearance of the commutator. Here he says he began from a need to create differential operators combinatorially. Differential operators can conventionally be both d/dx and d/dt, but the difference for Kauffman (as also for us) is profound because the (discrete) spatial displacement and the time step are different kinds of thing. The main steps are first to set up a discrete form of derivative. He envisages a sequence of "states of the world" S, S', S'' where S' is the next state to S in discrete time. We should interpret these as states of the hierarchy. A temporal derivative would then be $dS = S' - S$. For any abstract product of states (which he does not interpret at this stage but which is not the product in the hierarchy) it follows that

$$d(ST) = S'T' - ST$$
$$= (S' - S)T + S'(T' - T)$$
$$= (dS)T + S'dT.$$

Second, Kauffman restores Leibniz' rule by introducing an operator J by the rule

$$SJ = JS'.$$

Then he redefines a derivative by

$$DS = J\,dS.$$

Then

$$DS = J(S' - S) = SJ - JS = [S, J].$$

Since D is a commutator, it automatically satisfies Leibniz' rule.

Third, if we impose the idea that we are representing interactions (which we might call measurements) then a time jump appears: getting an interaction with a given result, and then doing the same with the interaction and the result in commuted order. These are different, and we treat the difference as a time click or clock tick.

Finally fourthly, one can study a simple case where the elements of the time series are commuting scalars. In this case, specialised to one variable, he considers the equation $[X, DX] = Jk$. This readily simplifies to $(X' - X)^2 = k$. This is a form of the diffusion equation. Here k is a constant scalar, and something essentially spatial has emerged. Kauffman makes this spatial step come about through a random walk.

One now asks about the order of progression of the argument. Kauffman is clearly concerned with, and possibly motivated by, the place of non-commutation in quantum mechanics. His first excursion into the algebraic structure behind fields[13] was about Maxwell's equations. In the same year[14] the same technique is applied to Dirac's equation.

Is it quantum mechanics where the argument begins, or is the appearance of the time tick and the complementary spatial step more profound and fundamental? In the above brief sketch we have inclined to the latter view. This presumption is important for our argument since the interrelation of clicks and steps also follows necessarily from the hierarchy construction. Hence we get an entrée into quantum mechanics that is acceptable given our principles, via Kauffman's commutators.

The hierarchy argument has to have levels if only to derive the coupling constants. Therefore there has to be some interpretation of the passage from one level to another. Our provision of this interpretation given above depended on the hierarchy's containing elements that transcended the first two-level separation. Looked at using realist language, these elements had the dual meaning (a) of two simple entities existing together and (b) of a single more complex one: they therefore had a place in both. Since the levels are constructed in a process, the appearance of the element is a click. Following Kauffman we call it a tick to emphasise that it appears before we have set up the mechanism for time with differential operators. In ordinary terms these two entities are the two spin states of the single particle. The spatial steps exist in the processes within one discriminately closed subset, and correspond with Kauffman's random walk. It follows that if the particle is the electron there has to be some sort of spatial separation as we see for example in the sodium lines or in the spin combinatorics that emerge from the Einstein-Rosen-Podolsky paradox and all the developments from that.

One naturally wants to go further using the differential operators that have appeared on the horizon to get fields. However only one step at a time: we do not take over mechanics in one piece. At risk of being repetitive we point out that the two-ity of entities we find to introduce spin does not entail that we have particles with macroscopic behaviour which can be used to connect with classical concepts such as fields and additive angular momentum until these have been defined. What we have in bare form at this stage and for the first time, is a spatial separation associated with the double entity. It comes as a shock to be told that one can have that without knowing how it will appear experimentally. It is important to realise that with this step we provide for the first time a meaning for spatial separation and therefore for space in any sense at all.

To see how it came about that spin was connected with rotation we need to understand why any particle with spin has a magnetic moment, and cannot be specified simply as a mechanical angular momentum with magnetic momentum added on. We can map out the route up to where the hierarchy gives the bare combinatorial necessities for electromagnetism, even though we may appear to be going head-over-heels into the continuum. Consider two test-particles which first exhibit mechanical behaviour and the change as they are said to exhibit electromagnetic behaviour. The mechanical behaviour has the motion of the test-particle in the same plane as any acceleration that is said to be due to a field. When the specification of the field is extended to include an acceleration at right-angles to that plane, we say we are dealing with electromagnetism. As soon as we treat the new acceleration as part of our theoretical description we are invited to consider that it may go either way corresponding to the sense of the charge. However what we have said so far covers both cases. This change corresponds with the inclusion of the elements in the second hierarchy level that link the two levels but do not appear in the first. They accordingly have an electromagnetic interpretation ultimately as differential operators in the Maxwell equations.

What this argument shows, albeit in classical language, is that in the hierarchy we go straight to electromagnetism at the basic level change. The purely mechanical angular momentum is a concession to the classical particle as a lump of matter cut up small and does not form part of particle physics in the modern sense.

In the rest of this chapter we shall comment on some other basic concepts that are in use in high energy physics in the light of what we have said about the two 'germs' — charge and spin.

Our consideration of mass raises two problems. Mass is like other descriptors in that each particle of a particular sort has the same mass but it is like the classical mass in taking a wide range of values. The fundamental aspect of the problems is the need to say how mass is generated in the particle picture. The supplementary aspect is to find why the particle masses have the particular values they do have. The Higgs picture is an attempt at the fundamental aspect. It seems unlikely that it will be able to contribute to the other aspects. The basic form of the Higgs picture is a bootstrap process whereby bosons build each other up by their being present together. In the nature of the case one of them has to start the ball rolling and that is the particle postulated by Higgs. That particle is bare of all attributes except mass simply because one has to start with something and one calls that mass. To make this process work there has to be a finite number of particles. Analogies are sometimes made with ions in solution where one of them acquires an 'effective mass' through its interactions with the ions surrounding it which is many times greater than its free mass. Unfortunately an analogy like this may lead readers to impute a similar sort of reality to the interactions in the boson case, and some of these models are quite bizarre. The whole point of the bootstrap is that it be abstracted from these classical-type models and mechanisms. These physical models already presume a classical-type meaning for "mass".

Perhaps the unclearness of the status of the bootstrap argument has been partly-responsible for the popular insistence on 'finding' the Higgs boson. People probably feel they could feel more sure about the bootstrap mechanism if such a particle could be found experimentally. There would then seem to be a clear 'yes' or 'no' to be got from the experimental search that would take the weight off the need for conceptual innovation. However, in our view, the origin of mass is a problem that is inescapably posed by particle physics and needs to be handled independently of the experimental reality of the Higgs boson. The problem existed for classical mechanics. Newton famously declined to pronounce on whether the mass that appears in his laws of motion could be seen as the source of the gravitational field.

Should the Higgs account of the origin of mass from a bootstrap build-up have a parallel in the hierarchy theory? The fine-structure constant reciprocal is the primary number to be given a physical interpretation and if we are to pursue the analogy with Higgs, it is necessary to ask whether there is anything that corresponds to the bootstrap process in its appearance. This question marks one of the points at which our combinatorial account and the conventional theory are most sharply at odds. Numbers that are supposed to represent the results of measurements are part of the real (or at least the rational) field. There is an important sense in which the fine-structure constant, in the way in which it emerges from our combinatorial theory, is not. Although the calculation gives a result expressed as a rational number, it makes no sense to suggest that it is just some member of the number field that might rather be some other nearby rational number, in the way that holds for experimental results.

We may feel that there ought to be some mechanism which would explain how the measurements give values that build up to the combinatorial value — presumably applying to one of the factors in $e^2/\hbar c$. In fact there cannot be any such thing and we are left recognising that it is merely our habit of thinking that makes us try to give conventionally built up values to these factors.

Turning now to exchange forces, interactions between particles in classical physics are mediated by fields which have a spatial distribution and a characteristic dependence on spatial separation. It became apparent that this kind of field was not adequate to explain the phenomena of high energies because of the difficulties in specifying its range. A threefold structure is quite different from the twofold interaction. 'Exchange' was introduced to replace the classical method of dealing with an interacting pair and to associate the combination with a new particle. (The three-body problem has not been solved.) Exchange was thus the basis for a new kind of dynamics. The idea is usually attributed to Yukawa, who imagined that there must be a new particle which introduced a new degree of freedom in the interaction of, say, two neutrons by its merely existing with them. The particle was spoken of as spending time with each of the original particles in turn. The word exchange was introduced. Yukawa was credited with having thus predicted the existence of the μ-meson (later realised to be the π-meson).

Exchange was always spoken of as a new kind of field. In fact it indicated a switch to a new kind of dynamics, and that fact was obscured by using the 'field' language. Thus one cannot understand the exchange as a contribution to the Hamiltonian due to a new moving particle. Yukawa mentions that earlier writers had prefigured the exchange idea. These included Heisenberg in particular, who had proposed to develop the analogy with chemical bonding by shared possession of an electron. In fact Heisenberg required something different from this because the chemical bonding is an ionic, and therefore static, coupling of particles — not a dynamic relationship. Yukawa says that nuclear forces are described by a scalar field U with a wave function that gives a static potential between two nucleons at a distance r proportional to $(e^{-\chi r})/r$, where the range of forces is given by $1/\chi$. Appeal is then made to the quantum-mechanical equivalence of fields and particles to give the new exchange particles.

This may have been the course of history, but the exchange idea is more profoundly new. We find 'exchange' in the bare combinatorics of a three-component manifold without the obligation to start with a spatial interpretation, but with the intention later to build spatiality into the picture given that starting point. That programme is what our hierarchy theory prescribes. The forces associated with exchange are supposed to have specific ranges, and the existence of these ranges is the ground for applying the field concept to them. In fact, however, it is not easy to be precise about these ranges: it was necessary to introduce an extra sophistication into the exchange theory to explain the finite range of the exchange forces. Moreover, as was pointed out early on for example by Yukawa, following Heisenberg, the 'saturation' of nuclear forces was hard to understand.

We have already mentioned the complex nature of the weak interaction. Sometimes strangeness is conserved; sometimes not. It seems the term weak interaction covers more than one thing. The hierarchy theory should have something to say on this point albeit tentative. We may expect strangeness to appear in somewhat the way $1/\alpha$ does. In the Parker-Rhodes construction the final stop occurs because the approximately 10^{39} dcss cannot be accommodated in the 256×256 places at that level. However 256×256 of them can. So we may say that the construction is possible in a small proportion of cases. Might not this proportion appear as the weak interaction? (Particularly as the form of the argument follows the lines used for other coupling constants.) The value given by this possibility is $1/(3+7+127+65536) = 1.52 \times 10^{-5}$. The measured value is 1.01×10^{-5} though there is some uncertainty about precisely what experimental value one should use.

In our rigorous calculation of the electromagnetic coupling, there are successive corrections to the bare value $1/137$. The first (and simplest) step is to imagine something cutting off the construction of levels beyond the Parker-Rhodes bound, and then, since that cannot happen naturally, to compensate for it. You have a $1/4$ chance of getting through level one. At the next stage the chance is $1/4 \times 1/8$. Then the chance for going further is $1/4 \times 1/8 \times 1/128$. Thus the combined probability (call it $1/e$) is 4096. This makes the Parker-Rhodes probability $1/137$ become $(1 - e)/137$ or $1/137.033$. The later work corrects this further to 137.0360.

In the light of this argument it seems that it may be better to treat the order-of-magnitude estimate above as a sort of upper bound, and that another figure may be got using this stepwise method. A rough shot is this. You have to choose matrices for mapping onto the dcss produced from 7 elements at the previous level, and this requires $2^7 - 1 = 127$ elements and $2^{127} - 1 \approx 1.7 \times 10^{39}$ places. However the number of 7-element sets contains cases where 1, 2 or 3 sums are all zero and therefore give rise to fewer than 127 elements. Thus we think of successive starts or attempts to pass the boundary: most of which require more places than there are available. The smallest case of 15 gives a multiplicity 32767, and these shots get through. The weak interaction strength that this argument gives is $(3 + 7 + 15 + 32767) \approx 3.05 \times 10^{-5}$.

Where then do we stand? The hierarchy group operations have a very direct physical or dynamical interpretation since they govern the creation of new combinations of elements that arise out of the ever-proceeding discrimination process. (It is true that this statement implies an unfamiliar use of the word 'dynamical' since a pre-existing space does not exist.) One expects to find a comparably simple interpretation of the operations in the **SU** groups, but the search turns out not to be simple at all. Ultimately one must find a path to the structure of space and time where transformations of axes are customarily represented by groups. One actually finds that the setting up of groups has been largely dictated by the shape of quantum mechanics, and that is why the symmetries have no simple mechanical interpretation but are abstract.

Perhaps it is the desire for a tangible way of understanding the group symmetries that has led to interest in invariances under C, P, T changes. Before exploring this history we look briefly at the overriding symmetries (presumably group symmetries of some sort) that particles exhibit, from the hierarchy point of view. 'Charge conjugation' is the view that the world should be unchanged by changing the sign of charge of any particle. This principle is thought of spatially as the requirement that when the axes are changed and the particle responds to a field by accelerating the other way, the laws of physics are the same. Clearly the physical picture is changed by changing the sign of the charge, but such a change can obviously be reversed as far as the space axes are concerned by taking the mirror image. This change is said to reverse the parity, and really does leave most of physics unchanged. It does however reverse the effect of changing the sign of the charge. Hence the combination of charge conjugation and parity can be used to preserve invariance, and this invariance is called CP symmetry.

In the hierarchy theory the change to a higher level provides more descriptive power as was described in Chapter 7. At level 2 there is enough descriptive power to provide a digital form of electromagnetism, and therefore changes in the sign of charge. If we refrain from using this freedom we can think in terms of mirror images, and can therefore say that CP symmetry is incorporated in the theory even though we cannot give the full meaning to coordinate rotations.

Certain recondite properties exhibited by particles break CP symmetry, but these all involve the weak interaction. We have only been able to give a sketchy picture of the weak interaction and that lies beyond the two first hierarchy levels. It is no surprise that basic new relationships requiring unfamiliar coordinate structures emerge with the weak interactions, and from our point of view there is no reason to think that interactions at all levels should fit the same coordinate picture. In the conventional view of symmetry-breaking time reversal may be expected to offset these breaks in CP symmetry. Whether they do really, has caused a lot of discussion — most of it obscure and not leading to convincing conclusions. The expression 'time reversal' is on the face of it nonsensical since classical language is being used, and one does not see time reversing itself universally. If some sort of local time is meant, then that needs to be fully introduced and explored. This difficulty is steered around by saying that 'time reversal' just means that wherever 't' occurs in the equations it has to be replaced by '$-t$'. What equations? There are no equations that uniquely characterise the weak interactions or indeed any of the rest of particle dynamics in the standard model.

We conclude that so far as symmetry breaking is a useful concept it is incorporated in the hierarchy theory, but its use to provide a general principle for understanding the particles is doubtful. When we turned to group structures that derived from quantum-mechanical foundations we were able to show that the more important theoretical ideas applicable to particle relationships that come from that source can be derived separately from the hierarchy. These group structures are abstract but still derive from quantum mechanics. We have taken the line that the philosophy that goes with the quantum mechanical mathematics is wrong and that we have to start again. How does that position affect what we derive regarding the numerical basis of the standard model from the group structures? The answer must be that the abstract structures can be used while replacing the conventional space and time picture with something consistent with the hierarchy algebra.

Chapter 9

Space, relativity and classical space

Contents

Our picture essentially discrete as is quantum theory. Based on the idea of discrimination. We look at the opposite (continuum) picture. A binary relation is essential to introduce space. This relation produced not by continuing the construction process but by repeating it. This introduces "pairs" of elements, thus departing from the construction that leads to levels and to an identification of particles using the numbers that characterise the levels. The pairs of elements make it possible to regard the second level basic structure as exhibiting three dimensionality.

In Chapter 3 we drew attention to the incompatibility that is usually thought to exist between quantum mechanics and relativity. Our theory starts with discrete sequences that represent the action of particles, and since discreteness is the territory of quantum mechanics we find ourselves on that side of the separation. We have already shown that our process theory, when the upwards construction is used exclusively, gives an accurate value of the fine-structure constant. In the last chapter we have shown that the horizontal construction yields the underlying structure of the quarks. A remaining task is to relate the process picture more closely with orthodox physics. That being so we have to give an account in this chapter of the other side of the picture with its continuum ideas, and we have to do that from the process point of view elaborated in Chapter 5. The strict continuum will necessarily be beyond our reach. Instead we present an approximation that requires the addition of new concepts to the process view.

Everything up to this late stage in this book has done without the concept of space using merely the method of sequentiality. We now review that method and recall the crucial point at which one of two alternatives was taken. These alternatives are the upward and transverse constructions described briefly in Chapter 7. We shall go into the details of the two constructions more fully here. The transverse construction came up in Chapter 8 in the construction of the quark model. Here we show that it also leads into the concept of space. The notions introduced in Chapter 4 are again essential and we recall their definitions briefly. A *discrimination operation* is one that determines whether two elements are different or not. This determination, and the corresponding binary operation takes different forms, as Chapter 5 shows. We define a *discrimination system* as a set of elements equipped with a discrimination operation. It is shown in Chapter 5 that a discrimination system necessarily has a set of levels.

If the two elements to be discriminated are at the same level the discrimination operation gives a third element if they are different and that element shows that the two have been compared and found different. If however they are identical it gives a non-element which we call a signal. The signal shows that they are identical. When the elements are at different levels the corresponding operation determines whether one element belongs to the other one in the sense of belonging familiar from set theory. Finally a discriminately closed subset (*dcs*) is a set of elements in which the discrimination between two essentially different ones is in the set. In the general flux of discrimination such a set will not be increased by randomly chosen discriminations between the elements. It should be mentioned that this definition of dcs includes the special case of a set consisting of only a single element together with its dual, since there are no essentially different elements. Correspondingly, in the skeleton (where discrimination is commutative) the definition includes the special case of a set of only one element. It is sometimes useful to use the skeleton because the skeleton throws away some of the structure in a systematic way in much the same way as homology theory in topology. What part of physics is being disregarded by using the skeleton will, of course, depend on the context.

The point at which the next step was unspecified has now to be discussed in more detail. When two elements come in, they will in due course discriminate to produce a new element. This gives a set of six elements in three dual pairs, or three elements if the distinction between new and old elements is ignored so as to produce the (commutative) skeleton. We shall see that we are forced to introduce a fundamental novelty that eventually leads to physical space. While we are restricted to the pure sequentiality of things going on in the construction or process, we may speak of the sequence as 'time', but it is a restricted time in which we are not allowed to dodge about. It is better simply to speak of 'order'. The elements form a totally ordered set as a result of the process. A partial order is needed to dodge about. Evidently we need to map our activities separately from the sequential order in which they were constructed. In fact we shall find that the new alternative in the construction process provides the binary relation that is essential to construct space. Space will be a special form of the manifold in which we can dodge about.

To see how a binary relation arises we observe that in the construction process a decision has always to be made between two possible procedures. When two elements come into play and produce a third, the three constitute a dcs and they are symmetrically related. Suppose now a new element comes into play. It is then possible for the system to go up a level. In Chapter 5 we assumed that it did eventually go up a level and that the exploitation of this extension produced a hierarchy and led to the calculation of the fine-structure constant. The rather obvious tactic of discriminating each of the three existing ones in turn was shown in Chapter 7 to allow an infinite regress and so is ruled out. Indeed any further discrimination with elements of the dcs cannot be the way ahead for the same reason. We are left with one possibility, and that is to leave the existing dcs as it is and start again. In due course a new element comes in and so the new ones will generate a new dcs. The new structure will then consist of pairs of elements, one from the original set and one from the new, say (a, b). From the method of construction the place of a in the succession must be earlier than that of b, a fact that will prove of importance later.

Algebraically the new structure is isomorphic to $S \times S$ where as usual S is the quadratic group. But it is more helpful to express it as two dcss, S, S' where every element of S' is later than every element of S. We write this as $S' > S$. As we explained in Chapter 2 and in Chapter 6 discrimination goes on all the time, successively producing quadratic groups (dcss). These dcss are partially ordered since, for instance, it is perfectly possible for two dcss, S, T to arise such that T has an element before any of S and another after any of S.

The new manifold that enables us to 'dodge about' was foreshadowed in the primitive ideas from which we began. The strong symmetry exhibited by the three in which no ordering is possible was taken to be the origin of three-dimensionality and therefore of a non-metrical description of space.

Because no ordering was possible, a new and distinct principle of ordering had to be invented, and that is what we have with our multiplicity of dcss. Of course there is a limit to the number of dcss that can be created in a discrete system. This is the Parker-Rhodes limit. Although there is no discrimination taking place between a and b, the process relates them in a way that is provided by the transverse construction. This binary relation is the first step towards the construction of space. The construction has resulted in an enlarged discrimination system whose elements are *pairs* of elements of the original discrimination system. We explore this a little more for the special case of the skeleton because it throws light on a curious feature of the original discrimination system. In the process account this system begins with a dcs of three elements. The connexion of this with the earlier Parker-Rhodes formulation in terms of bit-strings was stated in Chapter 5 by saying that the bit-strings were a representation of the algebra generated by the process. The three bit-strings $(0, 1)$, $(1, 0)$, $(1, 1)$ are pairs. They are just the three non-zero pairs of elements of the set $\{0, 1\}$ and this set is the trivial discrimination system consisting of a single element (and the signal, 0). The new construction has the same form except that now the strings are not bit-strings but strings (i.e. pairs) of elements of the original discrimination system. Let us call the elements of the strings the *underlying elements*. Then the underlying element of the original construction is a single one, whereas the underlying elements in the new construction consist of sets of three unorderably symmetric elements. The formalism just introduced is helpful in expressing the identification already given of time and space. The original construction gives the one dimension of time, and the enlarged system expresses the notion of space.

The foregoing discussion used the skeleton, and it is important to see what has been disregarded by doing that. The three spatial dimensions used in the skeleton have no sense of direction. Two elements a and a^* of the full treatment are reduced to one element of the skeleton, and the appearance of duality corresponds to the choice of positive or negative direction. Thus the full construction can express chirality as well as three-dimensionality. Way is then open to use the familiar language of electromagnetism.

The result of our construction should be compared with the "causal set" programme followed by some workers in quantum gravity[15][16]. Their model for space-time is a discrete set of "points" partially ordered by causal connectibility. The partial ordering is to provide ordered pairs of points (time-like separated). The structure "grows" by adding points: at each stage a new point is added to the immediate "future" of some existing point, or else to none when the new point is space-like separated from all the existing points. The set of points is locally finite, meaning that between two time-like separated points there can be only a finite number of others. It is clear that the process account produces an identical structure though without introducing the vague concept of causality, though we shall show later that it has the additional feature of three-dimensionality.

The profound change in thinking in the development of quantum physics is accompanied by an equally important change coming from relativity. A world in which changes in spatial position of a body have meaning only relatively to some other body, and in which there is a limitation imposed on the motion of bodies that is expressed by taking the finite velocity of signal transference as a universal principle happened early in the last century. It was a quite new world.

From our point of view the most significant thing in this new world was the appearance of finiteness as a physical principle through the finiteness of the velocity of light. To physicists who are not of our persuasion this finiteness is already fully expressed in the character of the velocity of light and nothing more needs to be said. If, however, as we do, we want to see the finiteness emerging at a more fundamental level, then we shall need to explain how relativity emerges from that change.

Accordingly we now turn to the derivation of relativity. There are two questions that we shall try to answer: (a) How is it that a theory like special relativity, discovered in a purely classical context, is also important in quantum physics? (b) Why are the inertial frames accorded such a privileged position in special relativity? The privileged treatment of inertial frames seems to contradict Newtonian mechanics because in that any frame can be considered inertial (if suitable force fields are allowed) but in special relativity the Lorentz transformation holds only between a special class of frames.

To answer these questions we have to investigate the notion of *frame*. We may try to interpret frame as the partially ordered set of points provided by our construction; but there is a difficulty. Frames, as usually understood, are indefinitely extensible, but the number of points that we can label is limited. This limit arises from our mathematics and is what special relativity is about. In bringing in the idea of a limit to a given frame we must contemplate more than one frame with one slipping past the other. Here we have introduced time: it is an extension of the time that we have already introduced. It is time with a sense of *duration*. By that we mean that we can halt the sliding and take a breath. 'Duration' is an essential, though mostly unacknowledged part, of what we mean by time. It is one of the new concepts we add to the process idea. It is obviously essential in a treatment of special relativity because the Lorentz transformation does not affect temporal order but does change durations.

In the same way as with time we need to distinguish between spatial separation and spatial distance when metrical ideas are introduced. We reiterate that when we do this it is not a matter of making the representation of space analogous with that of time. In the procedure of developing time beyond pure sequentiality (ordering) we have already had to import the idea of multiplicity of constructions that makes it possible to speak of space. When metrical ideas are introduced into time, we automatically have space.

The consequence of imposing a finite limit is associated in the conventional treatment with the physical measurement c. This is referred to as the velocity of light and is imagined as a thing of the same kind as the velocity v at which one frame slips past the other.

It is necessary to see how the finite limit is imagined physically, because, though the limit is imposed by our theory, its obtrusion into physical measurement is a thing that has strongly influenced current physical thinking and requires explanation at that level. We have to impose the finite limit that is conventionally associated with the physical measurement c. One frame is moved relatively to the one that is arbitrarily taken as a uniform standard. The limit is introduced in conventional treatments in a variety of ways that are all equivalent. In one, it is supposed that the moveable frame moves in steps that become ever shorter so that an infinite number of them never gets beyond one particular place on the first. One may picture what is going on, and secure the limiting process, by requiring that the size of each step is reduced successively in the ratio $\sqrt{(1 - v^2/c^2)}$ by saying that the frames have a relative velocity, v. Whatever way is used to introduce the limit, the factor $\sqrt{(1 - v^2/c^2)}$ plays an essential part and this function is a sufficient condition to provide the finiteness. We shall show how this factor arises.

There is one way of introducing the limit that apparently follows a different, contradictory, path. This is the formula for composition of velocities in the same direction:

$$v \star u = (v + u)/(1 + uv/c^2).$$

But applying this formula to find the increase in velocity owing to an extra velocity u being compounded with v gives

$$v \star u - v = u(1 - v^2/c^2)/(1 + uv/c^2)$$

and the familiar factor reappears. This function is a sufficient condition to provide the finiteness.

From our point of view both velocities v and c are only required as one (possibly the most concise) way of imposing the finiteness, but in the conventional view they form the basis of the whole of relativistic dynamics. A discussion of Lorentz invariance is usually part of the introduction of special relativity, and we have so far avoided it. The effect of Lorentz transformations is to preserve temporal ordering, and since we start from time as pure succession our spatial relationships automatically exhibit Lorentz invariance. Thus Lorentz invariance is automatic because nothing has been constructed at this stage to represent a deviation from it and neither will it be possible to deviate from it later. This position is very different from seeing it as a property that happens either to be satisfied or not. The constructive argument requires it to be there from the beginning as a central feature. This central feature will determine how the notion of duration that we have introduced, and that we shall elaborate below, brings with it that of spatial separation of things in addition to their being reached in a particular order. Both these principles have in effect been used in our explanation of the unusual properties usually attributed to the velocity of light.

In the abstract form of process, our most important argument has been to exhibit the way in which the finiteness of the process account replaces the finiteness that is regarded as due to the limiting velocity of light. Thus we explain why it is necessary to replace that basic set of dynamical concepts, and show how that can be done. In the process account spatial separation arises when we wish to make physical use of the purely sequential relation that the process provides which we have referred to as order. The relation between duration and sequential time has been neglected in the literature. W.H. Newton-Smith[17] speaks of the metrication of time, but his discussion is based on the prior existence of temporal lapse and is directed only at different non-equivalent descriptions of it. The most complete discussion of it is by Whitrow[18] some of which we shall use.

We shall take further our theory of the place of the velocity of light by using a construction that is based on the k-calculus due to Milne and Bondi and suggested but not exploited by Einstein in 1905 where you start with observers and signals. However you cannot even get the k-calculus from ordering alone. You need the two principles of spatial separation and duration introduced above. Being discrete, our theory is in a general sense quantum oriented and therefore will introduce non-local effects. If we set up an extension of the principles that permits the classical continuity we should not be surprised that that fails over a wide sphere. For these reasons it is a mistake to see the physics of the quantum and classical mechanics as two competing views of the world between which we have to choose, though of course they are two different methods of working. In particular the non-local effects of non-relativistic quantum theory do not infringe special relativity.

To introduce duration we must first go into more detail about the original ordering. It is tempting to use the word 'time' here though all we really have is a record of a sequence of steps in the construction process each labelled with an integer. The elements a, b, may have various meanings, but regardless of its meaning the element a occupies a definite place in the sequence. We shall use the ordinal numbers to denote these places. We shall call each place its *order-parameter*. We write the order-parameter of a as $\mathbf{n}$ when a occupies the nth place in the process. The heavy type notation is necessary because at this point the ordinals are just that. There is no question of doing arithmetic with them. Later, actual numbers will be introduced and we shall need to distinguish $\mathbf{n}$ from n. This is all there is to be said in the original construction but when two dcss are involved as in this chapter, the order parameter of an element will depend on the order parameters of its members.

The only requirement on this representation of order by ordinals is that of ordering. That is, the order parameter represented by $\mathbf{m}$ is earlier than that represented by $\mathbf{n}$ if and only if m is less than n. This requirement still leaves ambiguity for evidently the orders denoted by $\mathbf{1}$, $\mathbf{2}$, $\mathbf{3}$, $\mathbf{4}$ might equally be denoted by $\mathbf{8}$, $\mathbf{11}$, $\mathbf{13}$, $\mathbf{17}$. This would happen, for example if the four dcss quoted in the second list were, by discrimination, determined as the same as the first four, respectively. Two such orderings t, t' from the same dcs are essentially the same. It will be convenient to say t' is a *regraduation* of t if, for any $\mathbf{n'}$, $\mathbf{n}$, (order-parameters for the same element a) $\mathbf{n'} = \phi\mathbf{n}$ where ϕ is a monotonic increasing function. We can write that in the form $t' = \phi(t)$. If t' is a regraduation of t, then t must also be a regraduation of t'; that is, $t' = \phi^{-1}(t)$. Now the question is, why has t' been labelled in this way with spaces between $\mathbf{8}$ and $\mathbf{11}$? This question can only arise because there were other elements between that had not been recorded in the ordering. (The mathematician will note that although the inverse of a monotonically increasing function is usually another monotonically increasing function, for functions from and into the integers such inverse functions will not in general exist.)

Some further notation is required to denote these elements. Evidently what is essential for such a notation is that the elements in it should exhibit a *dense order*. That is, between any two there will always be another one. There are many ways in which this might be done, but the results will be essentially equivalent. The most convenient way is to define new times between $t = 1$ and $t = 2$. The regraduation states the need for two such times. A suitable notation for these is $1.1/3$, $1.2/3$ and so on for the other gaps. Nothing metrical is implied here. The meaning of $1.1/3$ is no more than "between **1** and **2** and earlier than $1.2/3$."

In this example the gaps occur only in one of the two order-parameters, but it is clear that the general case will have them in both. So the original notation for order-parameters in terms of the ordinals needs to be augmented to use the order of the rationals. Here again the heavy type is to emphasise that the rational field has *not* arisen — only a (potentially infinite) set of markers. The situation resembles that of Moh's scale of hardness.

We now turn from this elaboration of ordering to consider duration. To speak of duration is to speak of the lapse of time from a to b. This lapse is to be a *number* of some kind; not merely an ordinal. Call the lapse of time from a to b $f(a, b)$ so that $f(a, b)$ is some kind of number — a member of an ordered number field $\mathcal{F}$, say. Let T denote the densely ordered set of order-parameters, so that T is a set of dense order labelled by members of $\mathcal{Q}$, the rationals. Then f is a map: $T \times T \to \mathcal{F}$. We make two assumptions about durations that seem to us inherent in the notion:-

Assumption 1   The duration from a to b added (in $\mathcal{F}$) to that from b to c is the duration from a to c.

So for any elements a, b, c,

$$f(a, b) + f(b, c) = f(a, c).$$

Consequently, for *any* c ;

$$f(a, b) = f(a, c) - f(b, c).$$

Taking c as some fixed element, it follows that

$$f(a, b) = \psi(a) - \psi(b)$$

where $\psi(a) = f(a, c)$.

Next, fix b but suppose that a is replaced by a later element a'. Then

$$f(a', b) = \psi(a') - \psi(b)$$

and so

$$f(a,b) - f(a',b) = \psi(a) - \psi(a').$$

Assumption 2   If a' is later than a, then

$$f(a',b) < f(a,b).$$

So if a' is later than a, then $\psi(a') < \psi(a)$.

A first consequence of this conclusion from the two assumptions is that the number field $\mathcal{F}$ should have a dense order. There is some ambiguity here about what field should be chosen but it would be foreign to the discreteness of the process account to choose $\mathcal{F}$ to be the real field, even though that is the field taken for granted in the conventional treatment of the k-calculus. Indeed Hermann Weyl[19] has cogently argued against the identification of the intuitive passage of time with the real field.

We will choose $\mathcal{F}$ to be the rational field $\mathcal{Q}$. Then an important conclusion is:-

There is always a regraduation of order-parameters,
$t_a \rightarrow \phi(t_a)$ so that lapse of time is given by

$$f(a,b) = t_b - t_a.$$

In what follows we shall suppose this regraduation to have been carried out. Evidently it is not unique because a further transformation $t \rightarrow at + b$ where a and b are fixed rationals, will correspond to a change of the unit of duration. We shall refer to this notion of time as *duration-time* or *d-time*, to distinguish it from another that will appear shortly.

With this concept of duration established we can go to the notion of spatial distance. We shall find that this notion is bound up with duration, so that the way to explain it is via special relativity. This explanation contrasts with the usual treatment in which a three-dimensional Euclidean space is assumed from the beginning. We return to the situation described above of two dcss, and thus to pairs (p, q) of elements, one from each. Two such elements p, q are related by the process. By this we mean that b is determined by a but we do not write this as $b = F(a)$ because the elements a, b have algebraic properties exemplified both by discrimination and by d-times. We make this explicit by writing $b = F(t, a)$, where t is the order-parameter of a, together with $t' = f(t)$ where t' is the order-parameter of b.

There is an important difference, however, between our abstract language and the experimental. In the latter the idea of our discussion of frames as the necessary step into any spatial picture leads to a limiting velocity of light. This is a physical fact, but we derive it, so that no separate principle of maximal light speed is needed. In the usual idealization made to get to special relativity the photon is given the whole gamut of Newtonian particle-like properties. Which of these properties remains possible in special relativity can be determined by our deductive argument. For example the mass of the photon must be taken as zero. Our argument requires this because the photon has been introduced before any concept of mass.

The next step is to consider a further element, c say which stands to b in the process as b stands to a, so that $c = G(t', b)$, $u'' = g(t')$ where u'' is the order-parameter of c. Of course there is no reason to expect F, G to be the same function. Consider the special case in which c belongs to the same discrimination system as a. There are now two dcss with respect to which times are determined, so we use a prime to distinguish the two and have, for the d-times

$$t' = f(t), \qquad \text{and} \quad u = g(t')$$

where we have now written u for u'', the d-time at c, so as to emphasise that c is in the same dcs as a but later in the order.

Here the functions F, G are permutations of the elements of the dcss at each particular stage, whilst the functions f, g are monotonic increasing functions. We begin by considering the order-parameters alone.

If f, g are the same function then the two times t, t' will be called equivalent. This definition is based on a corresponding one described in terms of "observers" by Whitrow[20], which is a reworking of a somewhat different but equivalent definition by Milne. Strictly speaking one should allow for the ambiguity $t \to pt + q$ noted above, but that is unimportant. What is important, however, is a condition implied by the notion of equivalence that has just been defined. An equivalence should be a transitive relation. It turns out however that it will only be transitive if all the functions commute: that is $f(g(x)) = g(f(x))$ for any two. To see this, suppose there are parameters t, u related to t' in the second dcs so that the equations above hold, with $f = g$. Further suppose two more parameters v, w related to v' in the second dcs, so that

$$v' = f(v), \quad w = f(v').$$

Finally suppose t', v' to be related to x'' in a third dcs so that

$$x'' = g(t'), \quad v' = g(x'').$$

Then

$$x'' = g\,[\,f(t)\,], \qquad \text{whereas} \quad w = f(v') = f\,[\,g(x'')\,].$$

So the transitivity condition, that these should be the same function, requires $gf = fg$. This condition on a set of functions, has been discussed by Whitrow (*loc. cit.*).

Whitrow's paper proves that each function of the set has the form

$$f(x) \;=\; P[\,k\,P^{-1}(x)\,]$$

where P is a monotonic increasing function, the same for all the functions f, and k is a constant, a rational number (for us) that corresponds to the particular dcs. At this stage we must digress about the nature of Whitrow's proof. In his paper he naturally assumes all the functions to be defined over the field of real numbers, and he is therefore in a position to carry out his proof using the calculus. As we have already explained the real field is not available for use as it is quite at variance with our digital approach. Usually the rational field is a sufficiently good approximation to the real field, in the obvious way. That ceases to be the case, however, in questions like this one about limitations on the possible forms of functions. This is because functions may be defined over the rational field such as by taking $f(x) = something$ if the denominator of x is even and *something else* if it is odd. Hence it is no use trying to prove Whitrow's result over the rationals. However common sense shows a way out. There are two steps. First, in the theory as we have it, including the rationals, we can define *virtual* elements rather as geometers define points at infinity in Euclidean geometry, and we can ensure that these virtual elements play just the same rôle as the real numbers which are irrational. Then over this virtually extended field we can distinguish two kinds of function:- those whose definitions automatically go over into the virtual elements (and these we call *reasonable* functions) and those that do not (the *unreasonable* ones). Then Whitrow's result can be taken over in the form that every reasonable function of the set has the form given above, $f(x) \;=\; P[\,k\,P^{-1}(x)\,]$.

Such a set of dcss, in which the functions relating the orders of any two commute, will be called an *equivalence*. Each such equivalence is characterised by its function P, and this function is fixed because the ordering has been specified as the particular one for which duration is given by the difference of parameters. If however we remove, for the time being, this duration requirement then the above result shows that any two equivalences can be made the same by a regraduation. For if $x' = f(x) \;=\; P[\,k\,P^{-1}(x)\,]$, and if we apply the regraduation $x = P(y)$, which is permitted since P is monotonic increasing, then $f(x) = P(y') = P[k\,P^{-1}(y)] = P(k\,y)$, so that (using the monotonicity of P again) $y' = k\,y$. Let us call this new ordering time *privileged time* or *p-time*, in contrast with the earlier 'duration'. It will be noticed that this ordering strongly resembles the time in classical mechanics, but we are not quite there yet.

We can use these results to address the problem of preferred inertial frames. We have already used the binary character of our extended construction to introduce spatial separation. We now go on to show how spatial distance arises in the privileged frame. In this frame, the equations

$$t' = f(t), \qquad u = f(t')$$

become

$$t' = k\,t, \qquad u = k\,t'.$$

These formulae are familiar in the construction called the k-calculus due to Milne and Bondi, and suggested but not exploited by Einstein. There they are used to define a new, conventional time T for t', and a spatial distance X, by

$$T = 1/2\,(t + u)$$
$$X = 1/2\,c\,(u - t),$$

where c is the speed of light. The one-dimensional character of this definition is a blemish which Milne (for example) seeks to remove by a somewhat unsatisfactory later argument. For us, there is no puzzle since we have been able to postpone treatment of the general functions until we have dealt with the time-functions f, g. In the k-calculus this definition is not regarded as an introduction of space since the existence of space is already assumed in importing the physical ideas of sending and receiving light signals. Neither is it the first introduction of space by us because spatial separation appeared when we set aside the first dcs and formed a second one. What it does do is to introduce one-dimensional spatial distance.

Since it follows from expressions for t' and u that $u = k^2 t$, the value of X/T is given by $X/T = c\,(k^2 - 1)/(k^2 + 1)$, and in the spatial interpretation this is interpreted as a velocity of separation. Now apply these definitions to the discussion of equivalence as a transitive relation. For this purpose it is more convenient to rewrite the formula in terms of "advanced and retarded times":

$$t = T - X/c, \qquad u = T + X/c.$$

There are now two conventional times and X's to be attached to x''. The first is from the first dcs, call these T and X. The others are from the second dcs, and these we write as T' and X'. Then we see at once that

$$T' - X'/c = k\,t = k\,(T - X/c)$$

and

$$T' + x'/c = (T + X/c)/k\,.$$

$$T' = 1/2\,(k + 1/k)\,[T - v\,X/c],$$
$$X' = 1/2\,(k + 1/k)\,[X - v\,T].$$

Here the value of v is as before. These are easily seen to be the usual Lorentz formulae, since $v/c = (k^2 - 1)/(k^2 + 1)$ so that

$$k^2 = (1 + v/c)/(1 - v/c)\,.$$

Then

$$1/2\,(k + 1/k) = 1/2[\,\sqrt{(1 + v/c)/(1 - v/c)} + \sqrt{(1 - v/c)/(1 + v/c)}\,]$$

and that is the ubiquitous $1/\sqrt{(1 - v^2/c^2)}$. Of course, the existence of the square root in all cases would require the use of the real field but in this case that is of no importance to us because the quantity thrown up by the theory is k not v, and k is rational.

We can now see the problem of preferred inertial frames in a new light. The example of the k-calculus shows that the equivalence with privileged time corresponds to an inertial frame. This in itself does not solve the problem because a different time-variable was needed for duration. But it does show the way to a solution. The existence of preferred inertial frames is an empirical fact that is logically the same as the empirical fact that in such a frame privileged time and duration-time are the same. This identity constitutes a solution because it correctly locates the existence of such frames at a logically primitive stage. The difficulty was that, to quote Einstein, the development of special relativity required us to start with "a frame of reference in which Newton's laws hold in their usual form". He might have required this of Maxwell's equations too. The puzzle is that Newton's laws, Maxwell's equations, and observers moving at constant speeds need to be brought in to get to such primitive things as space and time measurements. In the rest of this chapter, in which the discussion is in an inertial frame, we shall simply use 'time' for the two identical parameters.

So far the generalisation of the k-calculus has been only a minor one. We turn to the major part for which we return to the general functions F, G; $b = F(t, a)$; and so on. As we said before, at each particular t, the function F must be a permutation of the elements. To make this clearer, suppose one dcs to be labelled (123) where $3 = 1 \cdot 2$. The whole dcs will include 1^*, 2^* and 3^* but these need not be specified as they simply arise as duals. Further suppose the second dcs to be labelled (abc) where $c = a \cdot b$. Then the function F sets up a correspondence which might, for example, be $1 \rightarrow a$, $2 \rightarrow c^*$, $3 \rightarrow ac^* = b$. We can say, to shorten the discussion, that the second dcs is (132^*). We have gone into this in detail to illustrate that the permutation part F is more restricted than being any member of the permutation group on the six elements $\{1, 1^*, 2, 2^*, 3, 3^*\}$. That group has order $6! = 720$. The F's belong to a sub-group of order $6 \times 4 = 24$. There is no great mystery about this sub-group, for if (abc) is any member, so are (ab^*c^*), (a^*bc^*) and (a^*b^*c). Thus its members can be written down as any member of the permutation group of order 6 on 3 elements, with the appropriate assignments of duals so that the third element is the result of discriminating the first two.

We have just described F at one particular time. As the process develops, different permutations will correspond. If we had been able to remain with the original order, expressible by ordinal integers, we could have imagined a corresponding sequence of permutations. However the new construction that gives rise to spatial distance requires a dense time-variable, which we have taken as specified by the rational numbers. It is physically unacceptable, as well as mathematically impossible, for different permutations to arise for each different rational number. We need a structure with a corresponding denseness to describe the group operations. The only way to provide such a structure is to suppose that at each stage there is a definite ascertainable probability of any one of the 24 permutations. The generalised ergodic principle would imply that, without further information, each of these 24 probabilities will be 1/24, so that deviations from these values will show physical interactions.

The next problem is how to incorporate these probabilities in an operator F that will now have the property of denseness. This is not quite straightforward. We shall find that the obvious way forward is ruled out and we have to adopt another approach. The algebraic need for this exactly parallels a physical feature in the space of everyday experience. This is important in confirming the correctness of our construction of space. The obvious choice would be to write F as the sum of the 24 permutations, each with a coefficient equal to its probability. This choice will not do. It is easy to verify that the result of two such operations is another of the same form (with the coefficients adding to 1 as they should) but the two operators do not form a group because there are divisors of zero. We exhibit a simple example of this and then show that the structure of the group of permutations permits an alternative. Consider, as an example, the permutation (132^*). The effect of this on the basic dcs, (abc), is to produce (acb^*). The matrix form of this operation is:

$$\begin{bmatrix} a \\ b \\ c \end{bmatrix} \longrightarrow \begin{bmatrix} 1 & 0 & 0 \\ 0 & 0 & 1 \\ 0 & -1 & 0 \end{bmatrix} \begin{bmatrix} a \\ b \\ c \end{bmatrix} = \begin{bmatrix} a \\ c \\ -b \end{bmatrix},$$

where $-b$ has been written for b^*. This change in notation is intended to reflect a physical difference between the separation implied by the skeleton and that implied by the full structure. The three-dimensionality of the dcs (123) in the skeleton does not reflect any sense of direction. In the full structure there are four corresponding dcss: 123, 12^*3^*, 1^*23 and 1^*2^*3 which do incorporate such a sense. One could express this by saying that the space of the skeleton is the positive octant of that of the full structure: but a better way of putting it is to say that the full structure contains right- and left-handedness which was absent in the skeleton. Now consider the matrix form of the permutation (13^*2):-

$$Q = \begin{bmatrix} 1 & 0 & 0 \\ 0 & 0 & -1 \\ 0 & 1 & 0 \end{bmatrix}.$$

If the permutations produced by P and Q enter the sum of 24 with probabilities p, q, respectively, the corresponding matrix will be

$$pP + qQ = \begin{bmatrix} p+q & 0 & 0 \\ 0 & 0 & p-q \\ 0 & q-p & 0 \end{bmatrix}$$

with determinant $(p+q)(p-q)^2$. *i.e.* $(p-q)^2$. So if no other dcss are involved and if $p = q = 1/2$, this determinant vanishes, showing that the matrix is a divisor of zero. The reason why this introduction of probabilities will not serve is because p, q are the probabilities of particular dcss rather than of the functions producing them.

It is possible to find a way of introducing the probabilities of the F's because of the structure of the group of permutations. We have been describing the process in general terms so we begin with a mathematical description of the structural feature, though this also corresponds to a physical feature of the space of our everyday experience as we describe below. The matrices P, Q are orthogonal (that is $PP^t = 1$, where t denotes transpose). It will be found in the same way that the matrices have this property. Consequently we are dealing here with an orthogonal group. This is an indication that we are on track to reach a Euclidean space but more argument is needed. The group we have reached, $\mathbf{O}(3)$, is doubly connected and as a consequence has two-valued representations.

To explain this, recall the idea of a group representation. A group $G = [P, Q, R, \dots]$ has a representation (one-valued is taken for granted) $g = [p, q, r, \dots]$ when g is a group and to each member of G there corresponds a unique member of g (here denoted by the same letter) such that whenever $PQ = S$ in G then $pq = s$ is in g. Notice that several members of G can have the same member of g corresponding. In a two-valued representation the uniqueness of the p corresponding to P is replaced by there being two members of g corresponding to each member of G. From the process point of view we would see the group $\mathbf{O}(3)$ as not the basic entity for the description of the F's. The two-valued representation, often written as $\mathbf{SU2}$ is the more primitive one.

Before we go further into the details, we explain the characteristic of the ordinary space of our experience that corresponds to the double connectivity of $\mathbf{O}(3)$. We do this by describing a simple experiment. Cut a narrow strip of paper of length about twenty times its width, and make a mark at each end on the same side. Now hold one end and twist the other through a complete turn (four right-angles) so that the two marks still face you but the strip has a screw form. It is then possible to move the two ends about in space, always keeping the two marks facing you, so as to make the strip flat, though now it has to be part of a ring. Any attempt to remove the ring-like nature (always keeping the two marks facing you) will cause the screw to reappear. Now repeat the process but with two complete turns (eight right-angles) so that the screw form becomes more pronounced. Then it is possible to move the two ends, keeping the two marks facing you, in such a way that the screw vanishes and the strip returns to its original form. The double-connectedness of $\mathbf{O}(3)$ means that the double twist can be removed, whereas for a single twist this is not possible.

For $\mathbf{O}(3)$ a most convenient form of this two-valued representation is by the use of quaternion algebra, and we give a, b, c the character of quaternion units by defining a new binary relation between them: $a^2 = b^2 = c^2 = -1$, and $ab = -ba = c$ and so on. To avoid confusion we should mention that this is not the structure that was found above from the process build-up. There is no reason why it should be, as it is here introduced simply to exhibit the transformations. Now let q be the quaternion $q = (1+a)$. Then since $(1 + a)(1 - a) = 2$, the inverse of q is $1/2(1 - a)$. Moreover it can be seen at once that $q\,a\,q^{-1} = a$, $\,q\,b\,q^{-1} = c$, $\,q\,c\,q^{-1} = -b$. That is, q produces, by means of the operation $q(\ )q^{-1}$ exactly the same permutation as (132*). Similar considerations will hold for all the permutations. The two-valued character is shown by q and $-q$ giving the same transformation.

The alternative incorporation of probabilities is by adding the corresponding q's with coefficients. Call the sum Q. The effect of Q is then to be given by $Q(\ )Q^{-1}$. This new set of Q's does now form a group and is indeed isomorphic to the rotation group in three dimensions. The rational coefficients are proportional to the probabilities of a q occurring. These probabilities will be different from those introduced above for the various dcss. It is of interest to look at the difference in more detail in a particular example. Consider $Q = p\,(1+a)+q\,(1+b)$, where p, q, are two probabilities. Omitting certain normalizing factors we can calculate at once that

$$QaQ^{-1} = pa + mb - qc,$$
$$QbQ^{-1} = ma + qb + pc,$$
$$QcQ^{-1} = qa - pb + mc.$$

The terms involving p, q alone are just the ones that would have arisen in the earlier calculation, but the terms in $m = pq \,/\, (p+q)$ do not occur there.

It is clear from this example that the effect of a general Q on a, b, c is to give expressions of the form $x\,a+y\,b+z\,c$ where x, y, z are rationals (which may be negative). Thus the need to have denseness in measure of duration again produces a dense space but one in which the transformations of the space are orthogonal, that is to say — preserve $x^2+y^2+z^2$, — where x, y, z are the coefficients that have become attached to the elements a, b, c by successive transformations.

Putting this with the Lorentz transformation which preserves $c^2\,t^2 - x^2$, will give a combined one preserving $c^2\,t^2 - x^2 - y^2 - z^2$. We have shown that we do not need Milne's weak argument to get out of his one-dimensionality, and at the same time we have clothed the bare bones of 3-d space with rational numbers (and so in effect have got angles).

Our approach is a complete contrast with the usual "Platonic Receptacle" view of space as a 'given' into which anything can be put. We have shown that the process necessarily constructs a three-dimensional Euclidean space, but it does this by a surprising path. The concept of duration allows special relativity to be constructed, and the Euclidean space can then be reached. Thus the usual route, in which Euclidean geometry is assumed first and then connected with an assumed time is completely reversed. Such a reversal means that there is little point in comparing our notion of space and time with others, with one possible exception. This exception is provided by Kant. It is true that he begins his *Critique of Pure Reason* with a detailed discussion of how our *intuition* of objects in space comes about, though Anschauung, his term, has rather different overtones. We can bypass this discussion because we have abstracted from such with our more general notion of process, with elements coming into play.

Kant says that the conception of space cannot be derived from outward experience but is a necessary *a priori* foundation for such experience. He follows this with a discussion of time, with regard to what he calls inner experience. We would have taken the two topics in reverse order, but apart from that we can see a certain resemblance between our view and Kant's. For us, order, from which time is later constructed, is the sequential process itself, and needs no constructive elements like Kant's objects of experience. The same remarks apply to the derived separation from which we construct spatial distance.

Chapter 10

Perception

Contents

Summary of historical views. Changed view of perception in quantum physics and the 'observer/state' philosophy. The process theory requires a change; there is no stage of perception of a thing separate from its construction. Many constructions compose the perceived world. Space (as distinct from the sequentiality of time) arises from the multiplicity of constructions. The passive observer has no place. Such a view accords with that of Leibniz for whom the physical thing was the appearance of a collection of monads. The paranormal and pan-psychism have to be allowed as possible aspects of the world since they are not excluded by our principle.

A natural sequel to the construction of space and time of the last chapter is a discussion of our perception of objects in it. More generally, 'perception' is whatever we do to acquire knowledge of the world. Philosophical discussions about perception that can be related to science have been long and various. On one side there have been philosophers who, for various reasons, simply rejected perception as the primary source of sure knowledge. Plato, Descartes and Berkeley predominate here. For Plato perception was treacherous: it could not address what later became the problem of universals, with the corresponding need for concepts. For Descartes the mind had to play the prior part if knowledge was to be certain, while for Berkeley reality was mental anyway. In the opposing camp the Stoics rejected Plato's scepticism, believing that the deceptiveness of perception could be avoided with care. William of Occam argued that abstract knowledge presupposed perception, and the extreme member of this second group, Locke, considered that all knowledge, including ideas, came only from experience.

Some reconciliation of these opposing views came with Leibniz, Hume and Kant. We discuss the relevance of Leibniz' views in relation to perception below. Hume distinguishes 'impressions ' which come forcefully from perception from the corresponding "fainter images" of them — simple ideas. For Hume these simple ideas can be built up into complex ones. Finally Kant sees an outer world that provides the matter for sensations: but these sensations are ordered by the mind by the use of a priori concepts. The sensations are perceived, but not the things-in-themselves.

In the heyday of classical mechanics most physicists supposed that there was a world outside that the theory described in all possible detail. In effect there was no separation between the theory and the world; one perceived the world direct and therefore had no reason to worry about what was implied in 'perception'. The traditional philosophic doubts about this attitude did not worry most physicists. Quantum physics changed that. It prescribed a stylised form for perception which was part of what we have called 'the observer/state philosophy'. This philosophy is discussed in Chapter 11 where reasons are given for thinking that it can never lead to an understanding of the particle and has to be replaced. What is needed for this understanding is discussed in Chapter 3.

In the process theory we do not set up the mathematics and then separately apply it to the world: it is assumed from the beginning that our construction can be applied wherever we wish, and if we want to make a distinction between theory and what it applies to then that is something that has to be done deliberately and separately. It follows that the constructing mind is not something over and above the process. One will then naturally ask whether there can be other perceiving minds, because if not then does the idea of a perceiving mind have any meaning or use. We are into the inquiry that philosophers sometimes call the 'other minds problem'.

For us, minds, one or many, must follow from our process construction. A single construction may be said to require us to speak of one mind — or it may be one soul. The only way for us to introduce space or distance is by duplicating our initial construction. We must still have our process construction, but we have to allow that we must duplicate it so there will be a multiplicity of these constructions. Since there is no way of tinkering with the construction internally to provide the multiplicity, we have to start anew each time. In Chapter 9 this duplication is used to bring in extension and physical space. Without taking this step we are not able to speak of a particle as a thing that moves in space or takes part in everyday dynamics.

However, the duplication here appears as a wider and more fundamental necessity.

To assert that there can be independent constructions, whether or not we speak of them as produced by independent minds, requires more thought about the ways in which it is possible to combine primal structures (typified by our quadratic groups) in parallel with, but independently of, the hierarchy construction. Eventually, proceeding along this line, we shall expect to reach symbols that represent dynamical variables that have *coefficients* so that a metric is introduced. For the moment however, 'independent minds' is as near as we get to the 'independent observers' that are usually spoken of linked with the idea of perception, and that would do the constructing of the metrical world. The word 'observer' leads us astray because it would prevent our thinking of their acting together in various associations.

The iterative character of the hierarchy algebra comes as a surprising step because physical space is usually thought of as a platonic receptacle (see Chapter 2) into which new experimental information is continually fitted. That physical space should itself be subject to construction is something new. Further thought about the part played by the mind of the experimenter in the iteration makes the step more than surprising — indeed completely strange. The step may be rejected out-of-hand as being altogether outrageous. In particular the passive observer has to go. If we are to speak of an observer at all, then he has to appear with the construction and deconstruction in the iteration. There may be ways in which sense can be given to a community of experience and so some sort of public world, but all that is a lot further on.

If the world becomes known to us through a multiplicity of our single constructions, and if we regard each construction as centre for perception, then we may try to use the term 'observer' for it. However, (in view of the inappropriateness we have found with 'observer') it may be better to appeal to a philosophic tradition in which the term 'soul' is used. This change of fundamental term gives a new twist to our enquiry, and leads us to ask whether philosophers who speak of *souls* could be said to be embracing ideas like ours.

We have already encountered the outstanding figure of Leibniz (discussed in Chapter 3 for his monads and his combinatorics in the context of *relational space*). His philosophy was profoundly important for the modern scientific outlook and he did indeed speak of souls. The fundamental thesis of his *monadology* was to see the universe as made up of individual substances that were soul-like entities: — the monads. The monads were capable of perception though in an unusual way. They were "windowless" but each one mirrored the whole universe, which made interactions possible. For Leibniz nothing can exist that cannot be built up from this base. The monads are not spatially extended and therefore Leibniz regarded them as immaterial. However Leibniz does want to get to a material world and this he does by treating a physical object as the appearance of a *collection* of monads.

There are striking correspondences between what our process theory demands and what Leibniz' philosophy supplies. We commented above on the strangeness of our demand that a spatial picture can appear only when a multiplicity of constructions exists. It is therefore gratifying to find this property of the spatial picture an essential part of Leibniz' Weltanschauung in which the physical object arises as the *appearance of a collection* of monads.

A monad is that which has no parts: it is indivisible. Atoms were the elementary entities most talked about by Leibniz' contemporaries. But the monads were different. They were not spatial whereas atoms were the smallest units of extension. Space and time for Leibniz are both relational.

A monad's properties include all its relationships to all the other monads in the universe, and each monad is a mirror of the universe. Each human being constitutes a monad, though not all monads are human beings. Each monad follows a pre-programmed set of instructions peculiar to itself so that it knows what to do at each moment. Monads are centres of force: substance is force, while space, matter and motion are merely phenomenal.

It may seem that we are proposing a world of pan-psychism. In fact however we have no interest in attributing *degrees* of psyche or consciousness to different stages of development of souls. Insofar as such creatures have psyches what we say applies to them all.

However a different vista opens. It may be that since we must contemplate souls being associated in many different ways, it follows that we must be open to the possibility that a wide variety of sorts of spatial and perhaps temporal configurations may depend on just what these associations are in any given case. Indeed what we should usually call objective space and time may no longer be a good guide to what to expect. We may have to contemplate a connexion between spatial and temporal relationships and the souls. If this is so, paranormal phenomena may have to be admitted as normal rather than as outside rational discourse.

This book is not about the paranormal. However we are concerned with the construction of spatial relations from sequential process and must be aware of all the ways of doing that that our terms of reference lay open. One would hardly expect to find in the writings on parapsychology a view of space and time that commanded wide acceptance, but the evidence that some of the effects described by parapsychologists are real is very strong indeed. Moreover the formalized effects are only part of a vast range of experience that is known directly to most people much of which is anecdotal or traditional. There used to be — say as late as the time of World War II — a majority of scientists who wrote off all evidence of the paranormai as fanciful because scientific theory was overwhelmingly successful and excluded every such possibility.

Fortunately in this matter, and in many more, scientists are far less triumphalist now than they were then. One who was of undergraduate age then would hear it said that even if there remained problems to be solved we were now in a position with the wide spread of science to know pretty well how to go about solving them. Science had got the basic picture. A look at current cosmology, with such paradoxes as surround our understanding of the dark matter, shows a fluidity in our very basic concepts and how they fit together that may be exciting in the extreme but at the same time engenders a new humility.

Chapter 11

Current quantum physics, state/observer philosophy: complementarity

Contents

The state/observer philosophy fails to provide the fundamental place for discreteness of the particle that our theory requires. It is inconsistent as a basis for quantum theory. The numerical successes of quantum mechanics with spectra are combinatorial in nature — not a new dynamics. Process Theory begins with the identification of the fine-structure constant and others which derive from the process algebra. The discrete nature of energy characterises the quantum theory. A summary of quantum theory is given so that its place for the discrete can be set beside that on which 'process' is developed. Planck's original formulation can only be correct if energy = hν. The development of quantum theory from this beginning is traced through the Rayleigh-Jeans Radiation Formula and the Bohr model of the atom. Schrödinger's influence is described : he was dissatisfied with the discrete jumps of the Bohr theory and proposed continuous wavefunctions. A fundamental alternative was introduced by Heisenberg which reintroduced discreteness through a new vision of mechanics in which observable changes alone were to be admitted in theory construction. These principles are applied to the harmonic oscillator. Attempts to provide explanations of the resultant difficulties. Only Bohr's discussion of sufficient depth but even this proves inadequate.

It was natural early in the twentieth century to ask that since classical mechanics had to be amended by some initially *ad hoc* devices, a new theoretical background should be found that would have the universal inclusivity that the classical picture had been supposed to have, and yet would

141

show the inevitable changes as a natural part of a satisfactory replacement. During the twenties the *state/observer philosophy* was invented — supplemented by the idea of complementarity. These doctrines prevail to this day and are supposed to be a sufficient answer to all the basic questions. Later in this chapter we shall show that dependence on the state/observer philosophy is flawed, and in particular that it does not help in explaining the origin of the discrete particles.

The purpose of this chapter is a negative one. In it we argue that orthodox (non-relativistic) quantum mechanics is not a step on the way to discrete particles. Consequently to take high energy theory based on it as a pre-requisite would not be a way to understand the origin of such particles. Our argument against quantum mechanics is that the theory is irredeemably inconsistent.

The chapter falls into three parts. First we show that the origin of the theory commits us to inconsistencies, and then we show that also these cannot be removed at a later stage because of the measurement problem. Compared with these difficulties for the theory, the fact that it is non-relativistic and that it is not related to general relativity at all, hardly matters. Secondly, while accepting that we have not yet any detailed combinatorial theory corresponding to quantum mechanics, we sketch two possibilities that might, separately or together, form a basis for such a theory. Finally we turn to the commonly urged ways of explaining away the difficulties, show that none is satisfactory and conclude by describing Bohr's very profound attempts in that direction.

The difficulties of the theory have led to different 'interpretations', none of which has succeeded in circumventing the inconsistency. More promising has been the notion of 'reconstruction'[21]. Reconstruction takes place in three stages. The first stage is to formulate a set of physical principles; then the second stage is to give a mathematical formulation of these. Finally the ambition is to derive quantum mechanics from the reconstruction. This idea is not new. Paul Dirac in the fifties came to consider his textbook to lack a Chapter Zero from which the rest would be derivable (personal communication). The first two stages correspond in our approach to the process account and its algebraic representation. Our third stage does not derive quantum mechanics but is combinatorial.

It may well be asked in view of our radical changes how we explain why quantum mechanics is nevertheless a highly successful tool for certain purposes. There is a widespread view that the numerical successes of the quantum theory are so massive that one is impelled to accept that theory even if it has confusions. This view needs to be considered and we suggest the following answer. The numerical successes mostly come in the area of spectral lines. The numerical successes were already there before the quantum theory came along. What the quantum theory faced up to was the discreteness of these calculations which was a mystery in classical physics. The quantum theory introduced the unit h to provide the discreteness, and if one can accept this imposition as a legitimate piece of theorizing one can say one has an explanation. It may be said that the discreteness of this sort and the spectral lines are part and parcel of a single piece of theorising, but that would be wrong. The spectral lines are obviously a combinatorial phenomenon, whereas the appearance of h is obviously something different.

If we ignore this change of logical character we naturally get an inconsistent theory. The precision of the scheme of spectral lines does not justify the inclusion of h in it. We believe this to be true notwithstanding the surprising successes of some extensions of the theory. For example the Coulomb term in the Dirac equation for hydrogen may be modified to take account of the action of the "electron cloud" on itself, and then the new (non-linear) equation changes the spectrum by exactly the Lamb shift[22].

Of course for us the origin of h is *indeed* combinatorial, and though it is a far cry from the discrete character attributed to h in the standard theory to the combinatorial that we give it, we do not have to face a break in the nature of explanation.

It is often argued that there are other experiments unrelated to spectroscopy that are well described by quantum mechanics. The discrete values of the spectral lines that correspond exactly with the mathematical scheme is different from the question of the explanatory power of the theory whatever one decides about its incoherence. The two-slit experiment is the clearest example. That is the one in which space plays an essential part. It is often said to be the experiment that characterises quantum mechanics. We shall argue again here that quantum mechanics is only a classificatory tool, but in order to avoid complication in the main argument we relegate that discussion to a Note at the end of the chapter.

Another problem is important for us : it is to relate high energy particle theory to process. That was attempted in Chapter 8. By contrast with quantum mechanics the process account introduces discreteness at the beginning instead of producing it by the introduction of Planck's constant h, as is shown by our ability to calculate an accurate value for $\hbar/e^2$. We emphasise that we are not here concerned to replace quantum mechanics as a highly efficient method of calculation, only to initiate an understanding of how it is related to the discrete process.

We begin by highlighting some peculiar but related properties of quantum mechanics. These have arisen in successive steps towards the current orthodox (non-relativistic) wave-mechanics. The first of the nine is the discrete nature of energy. One tends to see spectroscopy as the driving force behind early quantum mechanics. The wealth of nineteenth century measurements of sharp spectral lines accounts for this. But it is worth going to the turn of the century. Our intention is not to repeat a frequently told tale but to show these beginnings as an example of how the combinatorial nature of a result gets hidden away. At that time a major problem was to account for the way black-body radiation depended on temperature. An argument based on equipartition of energy gives the 1900 Rayleigh-Jeans law that the energy density per unit change of frequency is

$$U(\nu) = (8\pi k/c^3)\nu^2 T,$$

where k is Boltzmann's constant and T the (absolute) temperature. The Rayleigh-Jeans law cannot be right since it gives an infinite total energy, but it agrees well with experiment for low frequencies. Wien had shown in 1893 that the law must be of the general form

$$U(\nu) = k\nu^3 F(\nu/T)$$

for some function F. The Rayleigh-Jeans formula comes by taking $F(x) = 1/x$. Wien's measurements led him to assume instead that $F(x) = be^{-bx}$ where b is a constant and so to

$$U(\nu) = bk\nu^3 e^{-b\nu/T}.$$

It turned out that there was a value for b that gave good results for high energies but not for low. It was between these two formulae that Planck proposed an interpolation, taking

$$F(x) = b/(e^{bx} - 1)$$

so that

$$U(\nu) = (bk\nu^3)/(e^{b\nu/T} - 1).$$

The constant bk was written h, now called Planck's constant : so finally

$$U(\nu) = h\nu^3/(e^{h\nu/kT} - 1).$$

Later, in 1900, Planck sought a basis for a particular frequency. He postulated a statistical distribution of 'packets' of energy, each of amount E.

The original idea was to let E tend to zero after the calculation. Boltzmann's principle gives a mean energy

$$[E] = \int E e^{-E/kT} \, dE$$

between limits 0 and infinity. If E were a continuous variable, this would give $[E] = kT$. Planck tried instead the assumption $E = na$ where n is an integer and a is a constant. Then

$$[E] = \sum n a e^{-na/kT} / \sum e^{-na/kT}.$$

The series is easily summed to give $[E] = a/(e^{a/kT} - 1)$. This agrees with Planck's original formula only if $a = h\nu$. This is the real beginning of the Old Quantum Theory. The 'packets' of energy have a minimum size and this size depends on frequency. We draw two conclusions from this : firstly it is an example of a continuum situation which has, on closer inspection, a combinatorial character. Secondly, the result of this has been achieved at the expense of a basic inconsistency in the theory. The radiation theory comes from Maxwell's equations in which everything, including energy, is continuous. This result is now combined with a discrete energy. An inconsistency in a physical theory is not such a devastating situation as in mathematics but this one will be found to reappear in different forms as the theory develops. It does not arise for us since process theory is discrete from the start. This way out of the troubles that beset quantum mechanics is less easy than it sounds. More work is needed to relate the largely successful methods of calculation to the process.

Another example of hidden discreteness is in the Debye theory of specific heats. Here Boltzmann's statistical mechanics gives values for specific heats that agree with experiment at high temperatures but are wrong at low. The corrected version results from using Planck's notion of the energy quantum. It agrees pretty well over the whole temperature range.

We turn to spectroscopy and the slightly different considerations thrown up in 1913 by the Bohr model of the atom. At this point a curious feature of quantum mechanics begins to emerge. More than other theories in physics, its development has been largely driven by its mathematical formulation. For this reason we cannot avoid going into the formalism in discussing the ideas. We confine attention to the simple case, the hydrogen atom, in which a single electron orbits a proton in circular orbits. Bohr supposed that there were a set of orbits in which, contrary to Maxwell's equations, it would not radiate. These different orbits had different energies and a transition of the electron from one to the other caused radiation of frequency determined by Planck's rule from the energy difference. One can see the continuing inconsistency here.

In this model no theory is provided for the jumps of the electron between orbits. These 'quantum jumps' persisted as the theory developed, and a later manifestation is the "collapse of the wave-function". None the less the calculation led to the Rydberg formula for the sharp lines in the hydrogen spectrum. The special orbits were determined by a second "quantum condition". This second condition states that the angular momentum in the orbit should be an integral multiple of $h/2\pi$ which is now written $\hbar$. One sees here the beginnings of a genuine theory emerging. The possibility of elliptic orbits suggested a more general quantum condition : $\int p \, dq = n \, h$ for each coordinate q and its conjugate momentum p. The original inconsistency is reproduced as something that stays with the theory in its development. This feature is that, notwithstanding the appearance of generality by introducing p's and q's, the process of quantization depends on a particular coordinate system. This feature persists in the textbooks where the rule of writing down the Schrödinger equation presupposes Cartesian coordinates. That is because $p \, dq$ is not an invariant though $\sum p_r \, dq^r$ would be.

The more general quantum condition could now be applied to the harmonic oscillator. The importance of this is not only that this is the simplest periodic system but that a wide range of complex systems can, by a normal mode analysis, be reduced to an ensemble of harmonic oscillators. The classical energy equation is

$$m \, (dx/dt)^2 + m \omega^2 x^2 \;=\; m a^2 \omega^2 \;=\; 2 E \,.$$

If $x = a \cos \omega t$ so that $p = -m a \omega \sin \omega t$ and $dq = -a \omega \sin \omega t \, dt$, then

$$\int p \, dq \;=\; \pi m a^2 \omega$$

and so $E = n \hbar \omega$.

A new theory began to appear with Heisenberg (later with Born and Jordan). He proposed to use only observable quantities. Thus for spectroscopy allowed orbits and so on were banished. Instead the aggregate of amplitudes and frequencies were to be used. The frequencies come as differences of a discrete set of terms leading to Heisenberg's use of elements labelled with the pairs of energies that gave rise to the lines — that is, an (infinite) matrix. When q, p, are such matrices it could be shown that the diagonal elements of $pq - qp$ were all $i \hbar$ and so it was postulated that $pq - qp = i \hbar I$ where I is the unit matrix. This whole approach was tidied up conceptually by Dirac in the same year. He took q, p, as abstract quantities ("q-numbers") which came in conjugate pairs satisfying $qp - pq = i\hbar$. Notice that the notion of pairs of variables which are conjugate, familiar in analytical mechanics and prominent already in the quantum rules, now plays a basic role. It later leads to complementarity. The appearance of i at this point did not startle most physicists, with the notable exception of Schrödinger. In Heisenberg's original infinite matrix formulation it arose from the usual technique of writing waves of frequency ν as sums of $\exp(2\pi i n \nu t)$ rather than in terms of real circular functions. In Dirac's abstract formulation this explanation is no longer available. As we shall see below this "anti-commutation rule" is how discreteness is now introduced into the theory, (the "New Quantum Theory"). Notice that there is no introduction of space and time in this formulation.

The next step is to specify the physical system being studied. In the old quantum theory this step was not needed because it simply applied "quantum rules" to a classical system. The virtue of the new formulation is that the inconsistency generated by applying quantum rules to a classical model which forbids them is banished, by dropping the Newtonian model. Instead the system is specified by a Hamiltonian written with p's and q's which are q-numbers : but this Hamiltonian is to be looked upon as only a naming procedure. That is to say, to write $H = p^2/2m + \frac{1}{2}m\omega^2 q^2$, with $pq - qp = i\hbar$ is to do no more than to say 'This system is the quantum harmonic oscillator'. This use of the Hamiltonian as a naming procedure is our seventh peculiar feature. It raises the question of whether this trick really removes the inconsistency.

A third step is to specify how to calculate the energy levels. Heisenberg *et al.* were able to show that these would have been the elements of the matrix H when it was in its diagonal form. This way of putting it was no longer available to Dirac when q and p were abstract q-numbers. Instead the condition was put into the algebraically equivalent form $Hs = Es$ where E is a real number (the energy), and s is a q-number, described as "the state of the system". This is the origin of the notion of 'state'. The state will have certain numerical properties. These are the observables. Here p, q, E, are observables, and $Hs = Es$ says that the state has energy E (or sometimes that no observation has been made on the system so that it has to be assumed that the state is one of energy E). This 'state-observable philosophy' is our eighth peculiar feature, and will be discussed fully later.

The solution for the harmonic oscillator now gives $E = (n + \frac{1}{2})\hbar\omega$. To see this follow Dirac in taking new variables $x = p(m\hbar\omega)^{-1/2}$,

$y = q\,(m\,\omega/\hbar)^{1/2}$. Then $y^2 + x^2 = 2\,H\,/\,\hbar\,\omega$ and $y\,x - x\,y = (q\,p - p\,q)/\hbar = i$. The observable $A = (x + i\,y)\,(x - i\,y) = x^2 + y^2 - 1$, so that the values a for which $A\,s = a\,s$ differ by one from those of $2\,E\,/\,\hbar\,\omega$. The factors of A later became known as creation and annihilation operators. Now $(x - i\,y)\,(x + i\,y) = x^2 + y^2 + 1 = A + 2$. Hence $(x - i\,y)\,A\,s = (A + 2)\,(x - i\,y)\,s$. Putting $(x - i\,y)\,s = t$ gives $(A + 2)\,t = (x - i\,y)\,A\,s = a\,(x - i\,y)\,s = a\,t$, so that if a is a value for A it is also one for $A + 2$, which implies that $a - 2$ is another value for A. For a real system $x^2 + y^2$ cannot have negative values, so that the descending chain can only terminate at zero and possible values of a are $0, 2, 4, \ldots$ This gives the result $E = (n + \frac{1}{2})\,\hbar\,\omega$.

Two remarks may be made about this result. First, considerable ingenuity had to be expended even to handle such a simple system. The hydrogen spectrum was successfully carried out by this technique, but for more complicated systems it would be very difficult. Secondly, the result differs from that given by the old quantum theory, $E = n\,\hbar\,\omega$. An additive constant makes no difference in finding energy differences, but the extra "zero point energy" was welcomed because such a term had been inserted *ad hoc* into the old theory by Planck and Nernst in statistical calculations.

An important development came in the following year in Schrödinger's set of six papers. He was uncomfortable with the quantum jumps in the Bohr atom. His strategy to avoid them was to replace the abstract state of the Dirac theory with a "wave function" ψ of position (and possibly time as well). He had been influenced by de Broglie's ideas. Schrödinger originally thought of ψ as a physically real field, akin to electromagnetism. For a single particle Schrödinger's approach suggests the rehabilitation of space and time which the Dirac theory had excluded. This is only partly correct. For n particles ψ is a function of position in the $3n$-dimensional phase space (and presumably so in the case $n = 1$). So ψ is a function of q — a real number. Schrödinger noticed that the quantum rule $q\,p - p\,q = i\,\hbar$ would be automatically satisfied if p were $\hbar\,/\,i\,\partial/\partial q$. Thus the q-number quality of p was replaced by that of being a differential operator. This approach did nothing to ease Schrödinger's discomfort over quantum jumps. They reappeared as 'collapse of the wave function'. But a side product was a very efficient way of solving periodic systems, which accounts for its great popularity. Taking the harmonic oscillator again and making Schrödinger's substitution into the Hamiltonian, the equations become (in terms of the variables x, y, introduced before)

$$d^2\psi\,/\,dy^2 + (a - y^2)\,\psi = 0$$

where $a = 2E/\hbar\omega$. The usefulness of the reformulation is that the values of a for which this has a finite solution tending to zero at infinity are already well-known. They are as before

$$a = 2n + 1$$

for $n = 0, 1, 2, \ldots$. The corresponding wave-functions are known in terms of Hermite polynomials,

$$\psi_n(y) = \exp(-\frac{1}{2}y^2) H_n(y).$$

In later developments $|\psi|^2$ suitably normalised was interpreted as a probability density (Born interpretation). In the ground state of the oscillator $a = 1$ and $|\psi|^2 = \exp(-y^2)$. The classical oscillator would be a particle in the range $(-1, +1)$. The Born probability of the quantum oscillator being in this range is

$$\int_{-1}^{1} \pi^{-1/2} \exp(-y^2)\, dy = 0.8427.$$

The imposition of this range introduces the 'tunnel effect' into quantum mechanics. The quantum particle can tunnel through the potential wall that holds the classical particle together. This effect diminishes as the energy level rises.

Since $y = q(m\omega/\hbar)^{1/2}$, the exponential in the ground state is $\exp(-m\omega q^2/\hbar)$. Let us take as a conventional measure of the 'spread' Δq of q the value for which $|\psi|^2$ falls to $1/e$ of its maximum value, so that $\Delta q = (\hbar/m\omega)^{1/2}$. Now consider the momentum p. In terms of the variables x, we had

$$x = p(m\hbar\omega)^{-1/2} \quad \text{and} \quad x^2 + y^2 = 2H/\hbar\omega.$$

Now the commutation rule $qp - pq = i\hbar$ could equally be satisfied by taking p as an ordinary number and q as $i\hbar\, d/dp$. Introducing this into the Hamiltonian results in just the same differential equation as before, and so in the ground state

$$\psi(x) = \exp(-1/2\, x^2) = \exp(-p^2/2m\hbar\omega).$$

The conventional spread in p is then $\Delta p = (m\hbar\omega)^{1/2}$, so that $\Delta q\, \Delta p = \hbar$. This is an example of Heisenberg's uncertainty relation. The equality here arises partly from the particular convention used in defining Δp and partly from the simplicity of the example. In general it is found that $\Delta q\, \Delta p \geq \frac{1}{2}\hbar$, and this inequality holds for any two conjugate variables. There is thus a limit to the accuracy with which position and momentum may have definite numerical values simultaneously.

The sudden appearance of probability deserves a few words. What was said about the epistemic character of 'statistical' and 'random' in Chapter 7 applies equally here. It is ruled out for the standard theory but it is possible for us because of our ultimate discreteness.

There were many more technical advances in quantum mechanics after 1926 but we have said enough to make the difficulties clear. To sum up the position we have reached the basic inconsistency produced by importing discreteness into a continuum theory was ignored for a long time but was eventually removed by Dirac's formulation but only at the expense of introducing states and observables. The difficulty of quantum jumps was also accepted as a necessary evil. In terms of states and observables these jumps correspond to 'the collapse of the wave function' and so to what is now called 'the measurement problem'.

Because our central argument implies that quantum mechanics is inevitably inconsistent the second part of this chapter deals with two possible combinatorial ways of rescuing it that have never been tried. The initial inconsistency arose through introducing discreteness into a continuum theory. The later development concluded with a very clear formulation in Dirac's book in terms of the state/observer dichotomy.

The possibility referred to above is this : rewrite Dirac's theory using only the rational field and without the state/observable distinction. The rational field will not support calculus but it should not be too difficult to construct a finite difference version. In some respects it might be easier in considering the delta function. The major difficulty is the state/observer trouble. Some would argue that this trouble is of no consequence since states have to be 'prepared' and this preparation is performed by an observable. This argument is wrong however because Dirac's formalism cannot be constructed without both states and observables.

In the process theory we advance from pure sequentiality to a pluralistic philosophy with spatial relations. In the algebraic structure, this results from a construction giving rise to pairs of elements. Time then becomes more than the mere possibility of labelling the progression of instants. This step is involved wherever pairs of conjugate variables are needed, and that means the whole of physics. For this reason the appearance of conjugate variables provides our gateway into physical theory. For example, such pairing of variables is essential in discussing the attempt to measure simultaneously both position and momentum, and so in introducing complementarity.

It is usually supposed that these arguments have a common dependence on the existence of the quantum of action, h. For us the existence of h is part of a complex argument at a basic level and by no means merely just the way we impose discreteness. The further details of this approach remain to be worked out.

In the final part of this chapter we turn to the most widely held arguments to solve the difficulties in the foundations of quantum mechanics and conclude that they fail. Then we discuss Bohr's profound attempt in the same direction.

These difficulties that we find in the foundations of quantum mechanics are well known. There are several arguments aimed at explaining them away. We now list the most widely held of these and show that they are fallacious.

Some attempts to explain the wave-function collapse

People have always asked 'what causes one particular value to be chosen as the result of measurement when it could have equally well been any one of an indefinite range of values?' because it is a puzzling part of the state/observer formulation of quantum theory. These answers attempt to explain or avoid the supposition that the choice is made in the observation process. This list is not exhaustive and we mention only a few arguments that have gained most credence.

Many worlds

We begin with this idea because whatever its manifold disadvantages, it is alone in taking the state/observer formulation with no modification at all.

Everett (Hugh Everett III) argued that there was no need to find a mechanism that would choose one value since the quantum theory allows an indefinite number of random possible next steps all equally valid, and one of these has to be chosen. On the assumption that there is a continuum theory underlying this process this choice is sometimes called 'the collapse of the wave-function'. One may see the choice as happening necessarily in some one — out of many — developments, as Everett does. However it is essential to realise that the choice having been made the range of possibilities no longer exists and there is no way they can be discussed as real.

Everett however is not content with making this perfectly logical point.

He thinks that these choices represent alternative paths that really exist even though they are completely inaccessible. He sees each determination of a new state as the next step in one particular development that is really there. Thus there are infinitely many developments from this point all of which have a reality and each of which exists in its own universe of which ours is only one. The same is true independently for all the other points. These are the 'many worlds'. This new turn has dangers. One may be led to think that the many worlds are *simultaneously real* in the sense that their existence can properly form the starting point for various scientific and philosophical conjectures. In fact such extensions of the picture are ruled out by Everett's starting point in which nothing can be known about the reasons for choice of the element responsible for the wave-function collapse. If there were reasons we could never know them.

The many worlds picture has had support in some influential quarters, even though the implausibility of the multiply infinite paths that really exist, not to speak of our having to coexist with a whole lot of Doppelgänger, has put off most people[23]. The many worlds idea comes into its own when people such as David Deutsch discuss the virtual reality provided by computer constructions. The possibility of modelling of physical theory by computer models has played a part in our thinking but we reject the idea that one could use it slavishly to follow wave-mechanics. For our part we can't help being astonished that a mere technical mathematical device used to introduce discreteness into continuum theory (the collapse of the wave-function) should have had the power to provoke such weird speculation. What faith in the literal interpretation of the details of a formalism!

Hidden variables

Hidden variables were put forward to fill in the absent causation of the sudden discontinuous change called the collapse of the wave-function. They were 'hidden' because they had to have no other observable consequences. The Schrödinger equation acquired an extra term (the "quantum potential") that allowed the particles it described to have definite positions and momenta. The quantum potential, however, was a non-local effect. It never seemed likely that such convenient variables would be convincingly formulated, but the idea gained very wide notoriety. That was because it seemed to many people that the quantum theory had to be saved from its break with the principle that no questions should be ruled out in the discussion of any theory. Einstein was foremost in this criticism.

The hidden variables approach was mainly associated with David Bohm, though he later abandoned it. Attempts to extend it into relativity or into QED have made some progress (Basil Hiley, personal communication).

Measurement theory

The state/observer philosophy asserts that the quantum discreteness is due to the fact that something becomes observed, or, alternatively, that it is due to the observer. This position is not in itself illogical but no account is ever given of how the change is effected. There used to be a vogue for what was called 'measurement theory' which attempted to find the explanation for the effect of observation, or of the observer, in the details of quantum mechanics itself. This attempt involved arguing in a circle, since the dependence of just those details on the effects due to the observer were what was under review. A logically feasible way out was suggested by von Neumann. He suggested that the measurement process could be analysed in a regress of steps which would finish in the brain, or the mind, of the observer. No one else to our knowledge has taken the state/observer philosophy seriously in this way.

The prevailing method for providing a comprehensible basis for the state/observer picture is due to Niels Bohr. To give the setting for his views we first present the state/observer philosophy in the most conventional and unexceptionable way so that we know that we are in no danger of setting up a man of straw. Bohr's axiomatic approach (see for example[24]) can be put like this :

An observable is anything that can be measured — energy, position, a component of angular momentum, and so forth. Every observable has a set of states, each state being represented by an algebraic function. With each state is associated a number that gives the result of a measurement of the observable. Consider an observable with N states, denoted by ϕ_1, ϕ_2, ... and corresponding measurements $a_1, a_2, \ldots, a_n$. The theory postulates that the result of a measurement must be one of the a-values — that is a_1, a_2, or a_n, *etc.* No other value is possible. Probabilities are now introduced, but the essential step of introducing discreteness and allowing it to arise only through measurement has already been taken. The introductions of wave-functions which incorporate things other than measurements in addition, then follows.

Presentations of the quantum theory that follow this pattern all use the device of separating physical things into two classes :- those which have to do with measurement or observation, and those which do not. These two classes have quite different mathematical representations. This device is fundamentally flawed because it is arbitrary whether we call a thing a measurement or not. (The bits of matter that make up a measuring device do not know they are supposed to behave differently from those that are not.)

Present-day physicists do not worry much about justification for the 'state/observer philosophy' — considering it something thoroughly worked out in the past. If required to justify this complacency they would most often fall back on the work of Bohr and the principle of complementarily. We now therefore give a critique of complementarity as Bohr's principal foundational idea. There are writers who think that there are straightforward ways to avoid going into the questions that complementarity raises, and we shall consider these. It will appear that, though we shall find complementarity flawed, it attempts to tackle the underlying conceptual problems, and the alternatives do not. Bohr struggled to the end of his life to understand and justify the device of separating state and observer. Bohr is credited with the remark that "truth and clarity are complementary", and Peierls, to whom we are grateful for the quotation, added that Bohr leaned heavily on the side of truth. All attempts at clarity about observation, in the sense of a brief and definite statement within the intellectual structure which we call quantum theory tend to run into the kind of difficulties that have occupied us at some length already, particularly in Chapter 3. For example statements on the subject by Dirac are brief and admirably definite. However, the more we strive for clarity in such theories as that of Dirac, the more we find the difficulties thrown into sharp relief, and our 'clarity' is one which is obtained only at the price of being prepared to live with an underlying muddle. This particular sort of muddle, where there is a gap in the explanation which everyone tries to pretend they are happy with, would be referred to as a 'mystery' in theology. Feynman took the line that one got into such muddles or mysteries because one asked the wrong questions, and that this showed that one had not been 'properly instructed' : had not appreciated, in fact, what things just had to be accepted, mysterious or not.

Bohr differed from his contemporaries in the mainstream of quantum physics in not being prepared to temporise with an incomplete understanding of the basic quantum situation. However if, as we argue, the classical view of measurement and the conventional view of the way measurement is done in quantum physics are ultimately incompatible, then the endlessness of Bohr's search for a reconciliation was inevitable, and the accusation of unclarity, beside the point.

In this chapter we ask : does Bohr's complementarity principle enable us to understand, and therefore see as necessary, the differences between the 'quantum object' (Bohr's phrase) and the *object* of classical physics? At the time Bohr wrote, these differences were usually ascribed to the uncertainty principle. For Bohr, much more explanation was needed. So far we agree with Bohr, but we shall conclude that even Bohr's profound critique did not issue in an explanation, and therefore that no completely satisfying understanding of the differences exists at present.

The doctrine called 'complementarity' is used to underpin the concept of observation in mainstream quantum theory. It starts from the relationship of the particle picture and the wave picture, and goes on to a more technically articulated relationship that is said to exist between certain pairs of dynamical variables. These are described as 'conjugate coordinates' in the specification and solution of any single dynamical problem. The complementarity doctrine asserts that there is an exclusivity in the setting-up of two such techniques or concepts at a given time, even though both are needed for the full understanding of the problem. Born[25] gives the following brief and authoritative account :

"The true philosophical import of the statistical interpretation ... consists in the recognition that the wave picture and the corpuscle picture are not mutually exclusive but are two complementary ways of considering the same process — a process whose accessibility to intuitive apprehension is never complete, but always subject to certain limitations given by the principle of uncertainty. The uncertainty relations which we have obtained simply by contrasting with one another the descriptions of a process in the language of waves and in that of corpuscles, may also be rigorously deduced from the formalism of quantum mechanics — as inexact inequalities, indeed : for instance between the coordinate Q and momentum P we have the relation

$$\delta Q \, \delta P \geq h / 4\pi .\text{"}$$

In this account Born is more definite than most expositors in that he makes the Heisenberg uncertainty relation depend upon the more fundamental complementarity of the wave and particle pictures ('corpuscle picture', as Born calls it). However he leaves unanswered the question 'in which order does deduction proceed?' (Which is fundamental and which derivative?) He says that the complementarity principle leads to an uncertainty relation, and asserts that that is confirmed by rigorous quantum mechanics. Of course this would be fine if the more rigorous treatment included a more rigorous treatment of the wave/particle duality, but the treatment of that topic that appears in the above quotation is all the justification of it that he provides. Born's readers are left chasing round and round, and never sure at what point they are meant to break into the argument. Other writers are, as a rule, less clear than Born on this matter.

The evident — though perhaps never consciously expressed — invitation in such discussions as Born's is that one should build up support for the quantum-mechanical approach as a whole by deriving a little from each of an array of principles of which complementarity is one. Others are the Pauli exclusion principle, the commutation relations, and of course the Heisenberg uncertainty. The reader naturally wants to know whether these are independently necessary, or, if not, which entails which and in what order, but he gets no answer.

One might argue that such mutual dependence on a variety of starting points happens in classical mechanics. There, if we ask for a definition of mass, we are referred to statements which presuppose that we already know what force and acceleration mean : and *vice versa*; and so on round and round in circles. Moreover the closure of the system works. Everyone who is trained in physics knows exactly how to apply the classical theory and what constitutes a proper argument within it. The miraculous-seeming quality of the coherence is what we tried in Chapter 2 to draw attention to with our introduction of the term "theory-language". In the classical case we indeed have a closure of the definitions, which justifies us in starting from any of many equivalent points in our deductive treatment of any problem. Moreover, in a theory-language, every piece legitimately contributes to the meaningfulness of the whole. The vital point however, in the case of the classical theory-language, is that principles which would be invoked in justifying any one piece would be consistent with those for all the rest, whereas in the case of the quantum theory, this consistency is just what is being called into question.

Bohr himself was not content to see complementarity as a sort of philosophical gloss on the wave-particle duality which might make the duality more acceptable to those who happened to like that sort of thing. He singlemindedly presented it as an autonomous principle separate from the more technical principles of the quantum theory, and requiring no justification backwards from the empirical success of the theory. On the contrary it was this principle that should carry the weight of the quantum-theoretical vision of the world.

The following short statement by Bohr himself appears in an essay entitled "Natural philosophy and human cultures"[26]. He writes

"Information regarding the behaviour of an atomic object obtained under definite experimental conditions may, however, according to a terminology often used in atomic physics, be adequately characterised as complementary, to any information about the same object obtained by some other experimental arrangement excluding the fulfilment of the first conditions. Although such kinds of information cannot be combined into a single picture by means of ordinary concepts, they represent indeed equally essential aspects of any knowledge of the object in question which can be obtained in this fashion."

This definition of complementarity makes use of several principles which Bohr considered established in current theory, or which he considered he had himself established. Bohr maintained first that all information of physical significance is obtained by an experimental arrangement. Therefore the units into which it is alone legitimate to analyse knowledge about the world of quantum objects are whole experimental procedures (the appropriate experimental arrangements being imagined separated from the rest of the physical surroundings of the experiment). This principle sounds arbitrary until we see it against the special operational circumstances of the quantum objects. It is part of what we mean by the term 'particle' in the classical way of thinking that there should automatically be the possibility of defining other particles in the neighbourhood of the first without making any special theoretical provision for their intrusion. If we could not assume this without question, we should not be able to use the dynamical variables with their usual meaning. Now in the quantum domain this assumption is consistently and as a matter of fundamental principle invalid. If we wish to refer to a new particle then in that domain we must specify a new, and usually much more complex, theoretical background capable of describing the combined system.

The reader may well say at this point that having established principles for quantum physics its proponents *do* proceed to use the dynamical variables with their usual meaning, and will probably ask what else could they possibly do? To propose a possible alternative is the main purpose of this book, and we have no reason to persuade anyone that any less radical approach will work. Bohr thought he had a way through and that the way to express this specifically quantum view was to stress the unity of observed entity and observing system, and indeed to insist that neither should be ascribed reality independently. Bohr's assimilation of the atomic object (to use Bohr's phrase) to the circumstances of its measurement is further emphasised by what he called the "quantum postulate". This, he says[27], "attributes to any atomic process an essential discontinuity or rather individuality, completely foreign to the classical theories, and symbolised by Planck's quantum of action." One sees from this quotation that Bohr saw the very discreteness or particularity of the quantum particle as something to be imagined quite differently from the way we imagine a classical particle. In particular, the automatic classical assumption about other particles just mentioned would presumably be an imaginative prop that Bohr would require us to renounce.

A second principle which contributes to Bohr's idea of complementarity concerns the inevitability of the classical macroscopic description and the classical dynamical concepts. This, Bohr argues, is the only possible way of talking about the world in any of its aspects, including the quantum aspect. This second principle goes much farther than the first in the way it asserts that change from the classical language is forever ruled out. Bohr was insistent on this strong prohibition. Any suggestion that we ought to be open to change at the basic level of the intuition of spatial events so as to get a new kind of description appropriate to the quantum object, seemed to him entirely fanciful. This position of Bohr's seems at first sight like that of the naive realist who does not question the prevalence of classical language because it has not occurred to him to do so. Of course these positions are poles apart.

A good deal has been written about the influence of idealist ways of thinking that may have made Bohr feel he was on the right track in insisting on the classical language as though it were a necessary form of thought, or at least a precondition for all thinking that could be labelled 'physics'. In particular Bohr may have seen an analogy between the part played by the classical language and the synthetic *a priori* place of space and time in the Kantian philosophy. Again the complementarity idea certainly recalls the antinomies of Kant and the inevitable opposition of pairs of concepts in the dialectic of later German idealist philosophers. Petersen[28] has discussed these and related points.

The last component that we always find in Bohr's statements of the complementarity principle is that of incompatibility. We already have the unity of the operations and the language that go to make up a measurement : we have the restriction on the scope of that language to that which is current in the classical understanding of the world : now we are to understand that there will typically be more than one such description required to present the essentials of any given quantal situation, and that two or more of these descriptions may consistently appear so that the provision of one will consistently prevent the provision of the rest. As Petersen puts it (*loc.cit.*) "the experimental arrangements that define elementary physical concepts are the same in quantum as in classical physics. For example, in both cases the concept of position refers to a coordinate system of rigid rulers and the momentum concept refers to a system of freely moving test-bodies. In classical physics these instruments can be used jointly to provide information about the object. In the quantum domain, however, the two types of instruments are mutually exclusive; one may use either a position instrument or a momentum instrument, but one cannot use both instruments together to study the object."

WHY NOT? It is very difficult even to imagine what it would be like to argue in favour of Petersen's assertion, let alone actually to produce the argument. What *sort* of thing *could* prevent one kind of instrument being used because of the presence and use of the other? Would it be like crossing the critical mass boundary in neutron emission so that bringing up the second instrument would cause an explosion? Or would there be consistently destructive interference? Or what? Again, would the argument be that it was the successful operation of the one instrument that must inhibit the operation of the other? And then, what would the mechanism of the interaction between the two be? It is obvious that if one restricts oneself to classical arguments then there is no reason why one should not, for example, construct measuring techniques which measure momentum and position and other dynamical variables as well in indefinitely complex relationship. Indeed it is notorious that, far from its being the case that simple dynamical variables force themselves on the attention of the experimenter, his ingenuity is always stretched by the need to provide experimental techniques that exhibit those conceptually simple properties of a system that theory demands, in isolation.

There is one argument which might be put forward to show that a form of complementarity is a necessary consequence of the conditions of measurement. It is simpler than any advanced by Bohr, and yet it has a sort of logic and ought to be mentioned if we are to claim that we have exhausted all arguments that support the current epistemological doctrine about observation. The argument runs : "Suppose that we first measure position. If we then measure momentum, this must be by observing interactions with some sort of 'test-particle'. However we don't have test-particles which are small compared with the particle being observed (as we always do classically). Hence the momentum measurement must disturb the position, and this is merely a fact about measurement." One has to admit that a sort of uncertainty and a sort of limitation on simultaneous measurement do follow from the assumption of identical particles of finite size as the smallest available test-particle. The objection to the argument is that it is circular. The finiteness of test-, and all other, particles is supposed to be a consequence of whatever theory we propose to understand the quantum world, and it is therefore not available to use in the argument to establish that theory.

The conclusion we reach is that Bohr was not successful in using complementarity as the deductive basis for the quantum theory.

Let us end with something more positive. Earlier, we tried to explain Bohr's stress on the unity of the whole experimental procedure that leads to knowledge about a quantum object by bringing in ideas which are naturally to the fore in high energy physics. A quantum measurement does not presuppose the potential existence of a background of related experimental results in the way that a classical measurement does. Even for such a (classically) simple case as the measurement of two momenta at contiguous points at high energies, we require a quite different experimental arrangement from what is needed for the single measurement, and one that is usually of a much greater order of complexity. To see that this difference as a matter of principle and not merely of practicality is to come near to saying, as Bohr does, that different ways of measuring imply that different things are being measured. Indeed it would be only a small step to build the whole quantum-theoretical concept of measurement round the application of this idea of Bohr's to momentum and spatial position. Then one would have the uncertainty principle, in all essentials.

This way of looking at quantum measurement might well have been sufficiently radical for Bohr (who always stressed that his quantum postulate was something quite new on the horizon of physics) but it would not have satisfied him because it cannot be made to follow from the classical concept of observation. However Bohr had blocked off what we have argued to be the logical way forward with recognition of the relational nature of space, by his insistence on the classical description being the only starting point. In this book we regard measurement from a specifically quantum point of view, and treat the classical case as less basic; and so at this point we part company with Bohr.

Note 1.   The two-slit experiment

To fix ideas we begin with a short statement of a standard text book account. The experiment is a quantum version of the corresponding optical experiment (sometimes called Young's experiment). In each version a beam is incident on a screen with two slits, either of which may be covered during the experiment, and a further receptor screen. In the optical case a beam of light is involved; in the quantum case an electron or neutron beam is most usual. The two cases are not really so different since a light beam consists of photons. If the slits are narrow enough and only one slit is open there is a diffraction pattern on the receptor screen. If both slits are open this is replaced by a typical interference pattern. Two important features here require explanation : (i) the interference pattern is not that got by superimposing the two diffraction patterns : (ii) it is possible to conduct the experiment so that only one particle passes through the apparatus at one time. The single particle makes a single mark on the screen. But over time the aggregate of these marks makes up the interference pattern — as for a beam.

In a standard treatment for electrons the effect on the screen with the slits open can be approximated by treating each slit as a source of a stream of free particles and then solving the cylindrically symmetric Schrödinger equation. This equation is separable, the time dependence is harmonic, and the distance dependence is a Bessel function of order zero. This function is not very different from the product of a cosine curve with a slowly decreasing function of distance. In the corresponding optical experiment the wave-equation comes directly from Maxwell's equations and the solution is a cosine so the two cases are very similar. In each case adding the two solutions gives the usual interference pattern, and (i) is explained by the pattern being given by the square of the wave-function. It is feature (ii) that will give more in the analysis that follows.

The first thing to be said about this account is that it is not a quantum-mechanical description. It can't be since the treatment is exactly analogous to the optical one that simply uses the wave-equation derived from Maxwell's equations. Schrödinger's original idea was to construct a theory of matter waves analogous to the classical theory. The equation used has the same form as Schrödinger's in this simple case but the meaning is different. The "wave-function" does not have to be normalised, its amplitude is important. This point has been argued at some length by Tomonaga in a neglected book[29]. The point is that the theory gives a description of the wave properties of matter but not the particle properties. The use of h as a constant does not mean that the theory is quantised because it appears only in the form m/h (with the overall dimensions of frequency) and neither m nor h is fixed by the theory. The difference between this view of the theory and the theory as done by Schrödinger's equation is clear when one considers two interacting particles. The Schrödinger equation now has six variables. Only if the interaction is small can it be split up into two of the de Broglie type with a perturbation. The "waves" of the Schrödinger equation are in six-dimensional phase space which casts doubt on the use of the corresponding single particle equation to describe waves in physical space (but not that of the de Broglie equation).

Schrödinger wrote to Einstein in 1926[30] thanking him for drawing his attention to de Broglie's thesis[31]. Oddly, Schrödinger does not refer to de Broglie in the first of his six papers "On quantization as a problem of proper values" but in the second[32] he does refer to de Broglie's "elegant investigations, to which I owe the inspiration for this work." The distinction between a classical wave theory of matter and quantum mechanics arose only slowly in Schrödinger's mind.

Secondly, just as the New Quantum Mechanics was devised with the successes of the old theory in spectroscopy in mind, so that it is not surprising that it does serve as an extremely efficient calculational tool, so here Schrödinger's or de Broglie's equation is devised to provide a description of de Broglie waves and it is not surprising that it gives the expected results. The technique for quantising the de Broglie equation went through several stages. The result was set out with the utmost clarity by Dirac[33].

Thirdly let us consider the question of the experiment with only one particle at a time. The treatment above cannot describe this since it is not quantised. Schrödinger tried to describe a single particle by means of non-dispersive wave-packets, but that failed. In the final quantised theory Dirac showed considerable foresight in 1930. Speaking of the special case of polarised light he says : "It is necessary to suppose a peculiar relationship to exist between the different states of polarisation in which, for instance, if a photon is in a state 0 it may be considered to be partly in state α and partly in the state $\alpha + \pi/2$." "Thus the individuality of the photon is preserved in all cases, but only at the expense of determinacy." Evidently the concept of the particle has been changed.

Chapter 12

Just six numbers

Contents
This chapter title is taken from the book by Martin Rees. The six numbers discussed by Rees are seen by him from the point of view of cosmology. Why should they appear in microphysics as well? The numbers are dimensionless and their origin unknown. Rees lists them. By using the process method of this book they can be calculated. One of the numbers is interpreted by Rees as a measure of the 'ripple' or deviation from isotropy.

The argument of this book is, then that the origin of discrete particles lies in the conflation of the classical notion of particle with the combinatorial structure inherent in tbe progressive construction of a universe. As well as its intrinsic plausibility, this has detailed support in two ways: the agreement of the algebraic structure with that of the quarks, and the numerical values of certain dimensionless constants, particularly the fine structure constant. This chapter returns to these constants.

"Just six numbers" is the title of a book by Martin Rees. Rees shows that there are six numbers which determine the scale and indeed much about the nature of things in physics, and whose values are not understood. We by contrast claim understanding of some of them for which we can calculate values. Since some of these numbers appear in our basic scheme, it is necessary to compare Rees' reasons for high-lighting them with our own. Rees argues their importance from a mainly cosmological position. He implies, without actually saying as much, that cosmology *by itself* already exhibits this numerical scheme and that we should have been confronted by them even had we not known the inherence of the six numbers in the

163

physics of particles. However Rees's position deepens the mystery of their apparently having two independent sources. Rees is content to display this mystery and even perhaps to tantalise us with it.

Our contribution to the understanding of this mystery has of course been to explain that the two sources are not independent. If they are not independent then one precedes and the other follows. The argument has a logical order. Our entire view, which appears in Chapters 8 and 9, is that the numbers are the primary reality and the provision of any space in which they can be "located" needs to be generated deliberately for that purpose. New principles were required to add onto the basic theory and to explain and justify the generation because it was a major innovation.

The numbers are dimensionless and the dimensionless character is an inevitable part of the theory that underlies them. If we have to take them to comprise the understanding of basic atomic and cosmological constants then though dimensionless they have to be ratios of sets of those. Obviously it would be quite wrong to think that all ratios of such sets are examples of our fundamental numbers. For we can go on forming dimensionless ratios indefinitely, and we could not add them all to our list. Does their being ratios tell us the whole story about their non-dimensional character? Are there other dimensionless ratios of physically important quantities that we would not want to let in to our list, and if so why is that? For example what about the mass ratio of nucleon and electron?

Rees's numbers are these :

1. N.   The ratio of the electrical to the gravitational force; of the order 10^{39}.

2. ϵ.   Usually called the fine-structure constant, of the order 0.0007.

3. Ω.   Specifies the relative importance of gravity and expansion energy.

4. λ.   A cosmic antigravity; controls galactic expansion.

5. Q.   Ratio of two fundamental energies representing the size of deviations from isotropy; of the order 10^{-5}.

6.   Number of dimensions of space; 3.

The reason why these numbers have these values is unknown. At this point our theory offers the chance of a change since we calculate the values of 1, 2, 5, and 6 with varying degrees of precision. We shall also consider those, like the cosmical constant, which we cannot incorporate into our theory. These calculations came from a generalised particle theory, and therefore if they can be obtained cosmologically then detailed examination of the relationship between the two approaches is needed.

We begin our discussion with Q. We have related it to the strength, or coupling constant, of the weak interaction since that is the one of the fundamental dimensionless numbers that has approximately the value quoted by Rees. The value he quotes is presumably that usually given for the coupling constant of the weak interaction though there is much discussion in the literature of just how to allot its value in the last two decimal places.

Rees maintains that the number 1/10000 is a measure of the degree of 'ripple' in the universe. If we accept that this number is also the weak coupling constant, we are faced with the astonishing conclusion that these measures are essentially the same thing. Thus Rees's number Q really gets us into the mysteries. It is the ratio of the 'ripple size' compared with the universe size. One may think at first that ripple size is arbitrary, depending on at what scale we happen to look. However Rees argues that this ripple measure is all-pervading and must have a fundamental significance. He says it must have been present at the big bang. There are grave logical difficulties in thinking of the big bang as an event in time as 'time' is usually presented in physical theory and at the same time thinking of temporal relationships as taking their familiar form at the big bang. It may therefore be best to take Rees' statement to mean just that the origin of Q must be sought outside our spatial and temporal framework.

Why should the weak coupling have anything to do with non-uniformity? It seems very unlikely that we should be able to fit the two ideas into a unifying theoretical framework. However our theory does provide a way of relating them, and this we now describe.

The full process theory of Chapter 5 has a simpler version called the skeleton. It was Parker-Rhodes who first pointed out that the construction of levels must reach a stop when the number of subsets (dcss as we now call them) exceeded the number of elements available onto which to map them. However this stop rule is a statistical artefact since it is supposed that there is an overwhelming probability of a given dcs scoring a 'hit'. However there is a chance that it may miss and the probability of that happening can be calculated. We argue that if the Parker-Rhodes cut-off exactly corresponded to nature then we should have a totally smooth and isotropic universe. Since it does not, we may hope that the numerical value of the lack of fit describes a sort of overall deviation from uniformity. The idea is an imaginative leap, but what is the alternative? It is agreed on all sides that small changes in the value of this constant would have very large effects.

Our original conjecture was simple. In a certain proportion of the developments (of the hierarchy) one would hit the PR (Parker-Rhodes) bound, but that 256^2 cases would get through. The physical argument for this was never properly spelt out because what seems to be needed is the ratio of times the 'arrow' would get through, to the number of times it would fail. This is a much smaller number. In fact this argument was crude, and a different approach is needed. We must postulate a mechanism to reduce the effective multiplicity to 256^2 in which case the argument for number becomes comparable with that for the fine-structure constant in its own right and not as a ratio. It seems we have to think that elements in the hierarchy are created when and only when the Parker-Rhodes mapping has been performed. It was therefore premature to introduce the ratio of probabilities here.

The interpretation of these multiplicities as scale constants requires that they act as the descriptions of elements that act together as sets with a numerical magnitude rather than as a series of separate constructions or 'arrows'. The assumption that the generation or construction process simply gave more new entities than there were labels for in the product space needs amendment by putting linear independence of elements central. From that point of view one imagines searches going on at the 7-level to see how many elements one may start with. Only one competes because the higher ones are well above the mapping limit. Evidently the arrows are a way to describe the attempts at creation using different numbers of passed elements.

Another topic which is always heard in discussions of the fundamental numbers is often referred to as the "anthropic principle". The anthropic argument suggests that not much attention need be directed to the values of the coupling constants since those arise from a complicated interaction between what our evolution as perceptive beings will permit us to observe, and what the evolution of the universe happens to have provided for us. If the universe and its characteristic fundamental numbers were not much as they are, it is said, it would probably not be a possible place for us to exist in, and we should not know about the numbers, or at least their values might be accidental. Since we might have quite different observational abilities we might come up with quite different numbers. Or so the argument goes.

Our view is that interest in the anthropic principle is often overstated since we are proposing a reason why the numbers have the values that they do. In discussing the anthropic principle we are invited to imagine changes in the values of the fundamental dimensionless numbers, and to treat them as disposable parameters which could affect our existence, while keeping the familiar dynamics much as we see it now. This is the most serious difficulty about the anthropic principle because, whatever view one takes of the origin of the numbers, they are part and parcel of the theory as a whole and tinkering about with them involves very wide consequences for that theory each of which has to be considered in meticulous detail. There is no background theoretical framework that persists unchanged.

It is noteworthy that Rees puts the dimension number of space dimensions as one of his unexplained numbers. Probably most people would deny that there was a problem about that number and say that anyone with a metre stick in a four cornered room would immediately satisfy themselves about its value. In fact they would only get to the right conclusion if they already knew how to use the word 'dimension'.

The first problem addressed by the writers was to explore what the word 'dimension' could mean. Our answer to this question was the very first step that led over decades to the process theory of Chapter 5 and indeed underlay the whole argument of this book. It also provided the answer to Rees' question about the 3-dimensionality' by seeing in the use of the term 'dimension' the unique symmetry that was illustrated in the quadratic group of three elements (with the unit element).

Much follows from our early concern with giving a meaning — any meaning — to the word 'dimension'. How can we discuss 'dimension' without knowing what the word means? The strange fact remains that such baseless discussion goes on all the time and seems to be the cause of a great deal of misdirected effort. It is indeed true that the use of 'dimension' in the context of 3-space has a clear meaning, but our quarrel is with the automatic assumption that we can put in any other number for the '3' and still get a viable theoretical concept. It is assumed that many-dimensional theories are a simple alternative to 3-D that can be put in place at will. Among the examples of this assumption in action we may give that of 'string theory'. This theory is widely castigated because it has been around for so long and absorbed a great amount of effort without getting any results, but we suggest that this is, at least partly, because it is flawed from the outset. It makes this same blunder of treating the dimension number as a mathematical choice to be made freely, with each choice in principle as good as any other.

We are not at present able to give any account of the numbers Ω and λ.

Chapter 13

Quantum or Gravity?

Contents

Quantum theory and relativity clash. Incompatibility of continuum and discreteness. Our theory of the particle and of the dimensionless numbers that shape the categories of particles, requires us to start from the quantum end to get to the desired synthesis.

By way of epilogue to this book we acknowledge that physicists will expect to see the radical arguments of the book give a simple outcome on the quantum/relativity battleground and indicate how we have satisfied that expectation.

Most physicists whose interest is not specially directed toward the difficulties in reconciling quantum theory and continuum physics nevertheless know that there are these two main divisions of effort in physics. They will also be aware that there is no easy passage between them. One aspect of this division that is not often remarked is the way quantum mechanics has been driven into successive contra-intuitive notions by the mathematics demanded by the experimental results. By contrast general relativity proceeds from the familiar ideas about space which are then expressed in mathematical form, though this has very sparse experimental results. It is usually supposed that there is a clash between the quantum theory and general relativity that has not been removed. We may also point out that even special relativity, being a continuum theory, presents the more general problem of incorporating the discrete. The failure to reconcile the respective principles of the two theories is often thought to show a basic failure in our present view of the world.

Let us look at the arguments usually given for that view. General relativity is a theory about gravity, and quantum theory gives no account of gravity. Moreover, it is said, many of the formulae that describe the world as general relativity sees it, give ridiculous results when applied to the quantum world, and suggest things like infinite energies. The idea of a force also poses problems. Quantum mechanics accounts for three of the four forces in physics; electromagnetic force, strong nuclear force and weak nuclear force; but not gravity. On the other hand, general relativity describes gravitational phenomena very well but dispenses altogether with the Newtonian notion of a gravitational force. Hence the equations of quantum mechanics cannot easily be united with those of relativity. Even such an apparently straightforward generalisation as so-called Einstein-Maxwell theory, in which Maxwell's equations are inserted into a general-relativity manifold fails to provide a satisfactory unification of electromagnetism and gravity. That is because of the way in which force has to be introduced.

Before we can explain the contribution our theory can make we must elaborate and clarify these general arguments. It is necessary to distinguish the special from the general theories of relativity. Special relativity depends crucially on the puzzling notion of an inertial frame. Even if this notion is taken for granted, the relation of special relativity and quantum mechanics is problematical. Quantum mechanics is taken as a great success and similar claims are made for its generalisation to the electro-weak theory. Paul Dirac, one of the main creators of QED, has emphasised its incompleteness which exists because it depends in its calculations on the smallness of a certain constant, h, which it cannot calculate. Moreover quantum mechanics which started out as a theory describing a single particle, is turned into a theory of a field of many particles. Properties of single particles can be extracted only by the mysterious second quantisation.

To understand general relativity it is important not to start from the need to introduce gravity. The notion of a Newtonian force does not enter general relativity at all. This is because the principle of equivalence (that all bodies at a point accelerate equally) allows the Newtonian force at any point to be abolished by a coordinate transformation. Rather one should start from the inertial frames on which special relativity depends and then replace them by the general frames of reference between which the coordinate transformations abolishing the gravitational force can operate. Test particles then travel along shortest distance paths (geodesics). The coordinate transformations at nearby points differ, and that gives rise to another field quantity to express this difference (the Riemann tensor). The dependence of this tensor on the presence of matter is then given by

Einstein's field equations. They introduce a constant that can, by looking at the weak Newtonian field approximation, be identified with the original gravitational constant, G.

The field equations are often seen as a contingent addition to the necessary geometrical framework just described. A different view was taken by Eddington who saw the contracted Riemann tensor as providing the definition of matter and so as a step toward quantum mechanics. However he soon saw the incompatibility of the two theories. His answer to this was to try to find a problem which was sufficiently simple to be solved in either theory so that the results could be compared. In this he was unsuccessful.

These accounts fail to go back to the basic fact that relativity is essentially a continuum theory whereas quantum theory is not. This is the real incompatibility as is clear from the way the theories are formulated. The basic notion of general relativity is a differential manifold. It is true that singularities that represent matter are allowed but the treatment is by excluding them in a precise way. By contrast the Hilbert space of quantum mechanics is just a computational device to hold the state–observer calculus. And as we said above the single particle formulation of the original theory has, in QED, to be quantised again.

In classical mechanics, the changes in any field (including the gravitational field) can be made indefinitely smaller by bringing the interacting bodies nearer. In refusing to allow this, quantum mechanics makes itself entirely different. It is true that commentators on quantum theory are bound to claim that their quantum entities can be juxtaposed indefinitely to describe the appearance of the classical world, but this is done at the expense of a fundamental muddle which is due to the way we all behave as though there is always the familiar classical spatial world to fall back and rely on really. We have fully explored this muddle in Chapter 11. The paradox means that the physical spaces generated by the two are irrevocably different.

So we have two theories that see the world in quite different ways. It is not that one must be right and the other wrong but that they describe different aspects or ways of working. Can one not live with that? Apparently deep dissatisfaction remains and people are only happy to have two distinct ways of working if they know of a more general scheme which dictates the part that each plays. Evidently no such scheme exists and that absence is the root of what most people feel to be the quantum theory/relativity incompatibility.

We consider that our theory does provide a satisfactory scheme that will subsume the incompatibles. If they are really not incompatible then one has inevitably to start from one end of the divide and advance to the other because if one starts from both ends and hopes they will meet somehow in the middle then one is accepting that the incompatibility is final. Since we start from a view of the origin of particles, rather than of fields, the start we are committed to has to be that of the quantum — and therefore particulate. Given that, we simply follow our introduction of numbers that are a progressively better approximation to the variables of conventional physics which appear in Chapter 10. To do that entails the complete theory of process given in Chapter 5, and therefore we could say that our whole theory arises from the need to unify quantum theory and relativity, even though the origin of the theory is more general than that.

Because our starting point is discrete, in Chapter 9 we only reach an approximation to special relativity. From this approximation it would be possible to follow a path like the conventional one to derive an approximation to general relativity, but that is not our main claim. It has often been remarked that the ratio 10^{38} between electromagnetic and gravitational coupling constants needs explanation partly because of its largeness. None has been forthcoming because explanation was sought only inside the theories.

For us, it is no mere awkward fact that gravitation stands apart from the other forces. The sense in which it stands apart and the sense in which all the forces are part of a unified picture are (together with the calculation of the value of the fine-structure constant) the central successes of our theory. There is no difference between the way in which the interactions of basic fields appear, except in the vast differences between their strengths. Those differences are calculated; including in particular the enormous difference between the electromagnetic and gravitational strengths. It is the same features in the relationship of relativity to quantum theory that are usually seen as disturbing that emerge as the most satisfactory aspects of our unification of the wide picture embracing both theories.

References

1. H. J. Ryser, *Combinatorial Mathematics*, Math. Assoc. of America,
   Wiley, 1963
2. G. Polya, *Introduction to Applied Combinatorial Mathematics*,
   ed. Reckenbach, Wiley, 1964
3. H. Weyl , *Philosophy of Mathematics and Natural Science*,
   Princeton, 1949
4. A. N. Whitehead, *Adventures of Ideas*, p.168,
   CUP, 1947
5. B. Russell, *History of Western Philosophy*, p.576,
   Allen and Unwin, 1946, (2nd.Ed.)
6. L. Couturat, "Recent work on the philosophy of Leibniz",
   in: *Leibniz*, (ed. Frankfurt), Notre Dame Press, reprint 1976
7. I. Newton, *Opticks*, 1790 (4th.Ed.), Dover reprint 1952
8. H. P. Noyes, *Bit String Physics*, p.449,
   World Scientific, Singapore, 2001
9. F. J. Dyson, *Phys. Rev.*, **85**, p.631, 1952
10. A. Meyer, *Phys. Rev. Letters*, **83**, pp.3751-4, 1999
11. R. B. Palmer, *et al.*, *Phys. Rev.*, **25** B(5), p.323, 1968
12. L. H. Kauffman, in: *Proceedings ANPA 24*, Cambridge, 2003,
    Proceedings ANPA 25, Cambridge, 2004, p.72, and a useful
    summary in: *Proceedings ANPA 28*, Cambridge, 2007, pp.45-72
13. L. H. Kauffman, H. P. Noyes, *Proc. Roy. Soc. (A)*, **492**,
    1966, pp.81-95
14. L. H. Kauffman, H. P. Noyes, *Physics Letters (A)*, **218**
    1966, p.139

15.   G. Brightwell, H. Dowker, R. Garcia, J. Henson, and R. Sorkin
       "General covariance and the problem of time in a discrete cosmolog
       in: *Proceedings ANPA 23 'Correlations'*, Cambridge, 2002

16.   H. F. Dowker, "Causal sets and the deep structure of space–time",
       in: *100 Years of Relativity, Space-time Structure and Beyond*
       World Scientific, 2002, p.445

17.   W. H. Newton-Smith, *The Structure of Time*, pp.151,162
       Kegan and Paul, 1980

18.   G. Whitrow, *The Natural Philosphy of Time*, pp.169-175
       Nelson, 1961

19.   H. Weyl, *Das Kontinuum*, (1917),
       Translation to English, by S. Pollard and T. Bole, Dover,
       1994, pp.169-175

20.   G. J. Whitrow, *Quart. J. Math*, **6**, 1935, p.250

21.   C. Rovelli, "Relational Quantum Mechanics",
       in: *Int. Jour. Theor. Physics*, **35**, 1996, pp.1639-78; and
       A. Grinbaum, "Reconstruction of Quantum Theory",
       in: *Brit. Jour. Phil. Sci.*, **58**, 2007, pp.307-408

22.   A. O. Barut and J. Kraus, *Fundamentals of Physics*, **13**,
       1983, pp.189-94

23.   A useful comparison of Bohm with Ghiradi, Rimini, and Weber, is in
       Brit. Jour. Phil. of Sci., **59**, 2008, p.353

24.   See, for example, *Encyclopedia Brittanica*, (15th.Ed.), 2002, **15**, p.1

25.   M. Born, *Atomic Physics* (3rd.Ed.), 1994, p.144

26.   N. Bohr, *Atomic Physics and Human Knowledge*, Wiley, 1957

27.   N. Bohr, *Atomic Theory and the Description of Nature*, (3rd.Ed
       CUP, 1934

28.   A. F. Petersen, *Quantum Physics and the Philosophical Traditio*
       MIT, 1966

29.   S. I. Tomonaga, *Quantum Mechanics*, Vol.2, (Trans. Koshiba),
       North Holland, 1966, pp.169-175

30.   K. Przibram, *Schrödinger – Planck – Einstein – Lorenz :
              Briefe zur Wellenmechanik*,   Springer, Vienna, 1963

31.   L. de Broglie, *Annales de Physiques (b)*, **3**, 1925, p.22

32.   E. Schrödinger, *Ann. der Phys.*, **79**, 1926, p.489

33.   P. A. M. Dirac, *Quantum Mechanics*, (1st.Ed.), OUP, 1930

Index

175